Electron-beam Welding:

Principles and Practice

Electron-beam Welding:

Principles and Practice

Editor: A. H. Meleka
Bristol Engine Division, Rolls-Royce Ltd.

Published for The Welding Institute by
McGRAW-HILL

London · New York · Toronto · Sydney · Johannesburg
Mexico · Panama · Düsseldorf

Published by McGRAW-HILL Publishing Company Limited
MAIDENHEAD · BERKSHIRE · ENGLAND
for The Welding Institute

07 094218 8

PRINTED AND BOUND IN GREAT BRITAIN

Preface

The last decade has seen the evolution and introduction of a number of truly novel manufacturing techniques. Reference can be made to Electrochemical Machining, Plasma Welding, Laser Drilling, High Energy Rate Forging, and Electron-beam Welding. It is probably true that more new methods have appeared during the last decade than during the previous fifty years. No one cause can account for this upsurge; but the acceptance of new methods by industry is bound to be related to advancement in product design, the introduction of new materials, or both. There are also economic factors that may operate in favour of new techniques, but this is more of the exception, particularly in the early stages of the evolution of a modern process.

It is difficult to measure the relative impact of the techniques mentioned above on the metalworking industry, but there is little doubt that Electron-beam Welding has found application in a wider spectrum of industries than any of the others. Further, EBW has influenced design philosophy to a considerable extent in a number of specific industries. What it has really done is to elevate fusion welding to the plane of the high integrity structural joint.

To achieve such a status, many difficulties had to be surmounted. The inventors had to translate their prototypes into reliable machines that can be offered within an acceptable commercial framework. A dialogue had to be established between the plant manufacturer and the potential user. Suitable applications had to be found and the selection was often based on only a limited knowledge of the capabilities (and limitations) of the technique. Many disappointments were experienced by the pioneering user and both faith and courage had to be blended with technological skill. It is my belief that this book will take the reader through this evolution leading him to a technique that is now a reliable and essential day-to-day production tool in many industries. A measure of the success of EBW can be gained from the fact that it is estimated that there are nearly 1000 machines in the world today; the majority are in the United States and the United Kingdom.

The book is written primarily for the user and potential user of EBW; indeed for any industry utilizing metals and alloys. Some aspects of the subject will be of interest to the student of Production Engineering, Welding Technology, and Metallurgy.

The book falls naturally into two divisions. The first few chapters provide an insight into the technique, its history, the fundamentals of beam generation and control, the thermal effects of an electron beam, the mechanism of deep penetration, and also an analysis of the basic features of an EBW plant.

The second half of the book is more directly concerned with application, outlining the principles of component design, metallurgical considerations both favourable and unfavourable, also production engineering matters including economics. There is a chapter concerned exclusively with applications; each application is derived from the author's own experience. This chapter illustrates the wide range to which the technique can be applied, from minute instrumentation components to large jet engine casings.

To edit a book based on the experience of some of the leading pioneers of EBW has been a source of pleasure to me. I am also honoured by participating with them in my capacity as a fellow-contributor (chapters 3, 4, and 6). I wish to record my appreciation to the following (and their organizations) for their co-operation and support:

R. Beadle, BSc, ARCS	Vickers Research Establishment (Chapter 2)
J. V. Birnie, BSc, AIM	Hawker Siddeley Dynamics Ltd —Electron Beam Division (Chapters 4 and 6)
M. E. Boston, MA (Cantab)	Cambridge Vacuum Engineering Co. Ltd (Chapters 1 and 8)
P. A. Einstein, MSc, MIEE	Vickers Research Establishment (Chapter 2)
M. J. Fletcher, BSc, PhD	GEC Power Engineering Ltd (Chapters 5 and 8)

R. N. Lower	formerly of Appleby and Ireland Ltd (Chapter 7)
J. K. Roberts, MWeldI	Rolls-Royce Ltd, Bristol Engine Division (Chapter 5)
D. V. Russell	Sciaky Electric Welding Machines Ltd (Chapter 4)
J. D. Russell, BSc, AIM, MWeldI	The Welding Institute (Chapter 5)
A. F. Taylor, BSc, AIM	UKAEA (Chapter 1)
H. White, AWeldI	Rolls-Royce Ltd, Aero Engine Division (Chapter 7)

All contributors pooled their experience in the Applications chapter (7).

My thanks are due also to Miss Mary Woozley of The Welding Institute for her efforts in co-ordinating the project and for her assistance in the editing. I am also greatly indebted to Mrs L. Skinner for typing the manuscript.

A. H. Meleka

Contents

1. History of electron-beam welding

Occasionally, an invention may come about by accident. More frequently, an invention is the culmination of a logical process in which a number of clearly defined prerequisites can be found. Firstly, a need must exist for a new process or a new material. Secondly, such a need must be clearly recognized by someone who is directly involved. The third, and perhaps the most important, requirement is that, whoever recognizes the need, the potential inventor must also be capable of supplying the solution. It is in this context that we seek to examine the history of electron-beam welding (EBW).

Frequently an invention has not one but two origins. This was certainly so with EBW and we shall be examining two distinct and quite different stories of invention. In one, the invention arrived as an answer to a clearly defined need, while in the other the invention was pursued for its own sake. Consideration of the circumstances which brought the invention to fruition, when examined in the light of the above analysis, may present a different, more realistic, and more interesting picture than by simply cataloguing the development of electron beams and their application to welding.*

An examination of the literature on EBW soon reveals that early papers on the subject refer exclusively to applications in the nuclear

* In dealing with the history of EBW, the authors of this chapter must refer, by necessity, to certain dates and events which indicate the relative contribution and priority of invention of the early workers in this field. Besides the technical aspects, there are naturally commercial matters such as patent rights involved. In this context, the authors, editor, and publishers of this book accept no responsibility for the accuracy of the details given in this chapter. The information here included was obtained either from already published work or direct from the early workers themselves.

industry and that the matter was associated with a certain class of materials used in that industry. From about 1950 onwards, there was an increasing interest in the use of refractory metals for nuclear applications. Materials were selected for such properties as their low neutron-capture cross-section, their high melting point, or their compatibility with uranic fuels. Selections included niobium, tantalum, zirconium, vanadium, beryllium, molybdenum, tungsten, and their alloys. These were the potential canning materials for the more advanced reactors under consideration at that time, but they could hardly be used unless satisfactory joining techniques could be developed. All these materials have one characteristic in common: they are extremely reactive, readily combining with oxygen and nitrogen at temperatures in excess of about 300°C. The impurity levels thus introduced have a marked effect on the corrosion properties, on the strength, and on the ductility of the metal. For this reason any operation such as welding which involves heating to high temperatures must be conducted in an inert atmosphere.

This requirement led naturally to the assumption that inert-gas tungsten-arc welding would be a suitable technique. Early attempts, however, showed that the standard process did not provide adequate shrouding by the inert gas, so the shrouding was extended and welding in vessels purged with inert gas was tried. Such techniques proved successful only in a limited number of applications. It was soon appreciated that satisfactory inert-gas purging could best be achieved by pre-evacuation of the vessel. Having reached this stage, i.e., the use of vacuum equipment, it was apparent that a vacuum provided less potential impurity than the best inert-gas atmospheres obtainable. However, as an arc cannot be maintained in a vacuum, the ideal method of welding refractory metals was yet to be found and some workers continued to look for other alternatives.

It had been known for many years that a beam of electrons accelerated in a vacuum would induce heating in the target upon which the beam impinges. It was probably a natural consequence therefore that engineers with background and experience in cathode-ray tube and vacuum technology should consider electron bombardment as a welding heat source. Thus, we have the right prerequisites for the invention. The ideal circumstances indeed existed in 1954 for Dr J. A. Stohr of the French Atomic Energy Commission, the CEA, to initiate his developments which led to the electron-beam welding technique. Stohr first made these developments public in November

1957 at the Fuel Elements Conference in Paris.[1] This must undoubtedly be the key date in the history of EBW since it was this publication which really originated the worldwide interest in the process.

It must be remembered that, at this stage in the history of the development of EBW, the main concern was to develop a process which would permit fusion welding in a vacuum. The early CEA unit incorporated a relatively simple electron gun normally operating at about 15kV, using electrostatic focusing, and with the work acting as the anode. A diagram of the unit is shown in Fig. 1.1. The

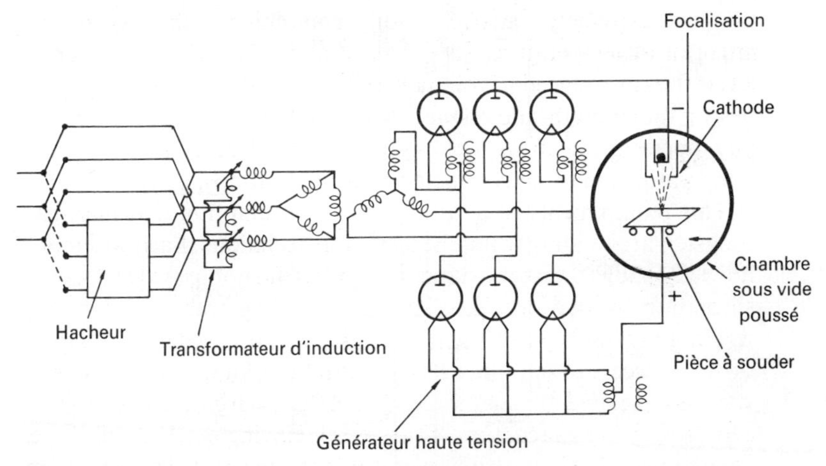

Fig. 1.1 Equipment developed by Stohr ('Schéma du circuit d'alimentation électrique', Stohr and Briola *Vacuum welding of metals.* IIW Annual Assembly, Vienna 1958.)

controls permitted variation in accelerating voltage, filament-heating current, and bias voltage for focusing. All of these parameters were interrelated and alteration of one involved adjustment of the controls of the others if their values were to be kept constant. The welds produced by this equipment were remarkably neat and uniform in external appearance, with a depth-to-width ratio twice as great as experienced with TIG-welding. Thus the main attraction of the process at this stage was merely to fusion weld refractory metals in vacuum, i.e., the purest 'atmosphere' available.

The nuclear requirements described above naturally existed in the United States also. Equipment similar to Stohr's was developed independently by W. L. Wyman at the Hanford Laboratories, Richland. However, the work was started a little later—in 1956—and was reported in February 1958.[2] A diagram of the unit is shown in Fig. 1.2 where it is seen to be almost identical to Stohr's equipment.

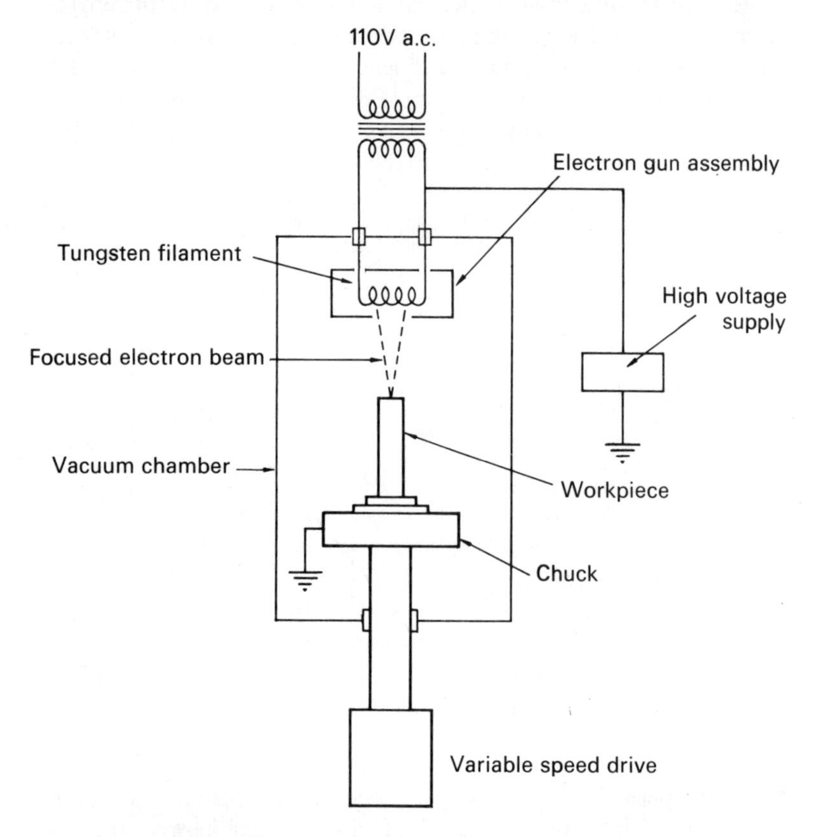

Fig. 1.2 Equipment developed by Wyman[2]

The accelerating voltage was limited to 15kV which was considered the upper limit of usability; higher voltages would generate hard X-rays capable of penetrating the vessel walls. Again, the equipment was developed as a means of fusion welding refractory metals, particularly Zircaloy-2, in a vacuum. Curiously enough, one material which Wyman reported as not being amenable to EBW was aluminium, whereas Stohr had successfully welded it.

No independent EBW development was reported in the United Kingdom. The first equipment available was produced by Edwards High Vacuum Ltd towards the end of 1958; this was manufactured under licence from the CEA. The equipment provided a maximum accelerating voltage of 20kV and a maximum beam current of 100mA. An added refinement compared with the original Stohr gun was the provision of a magnetic lens to give improved focusing. The first three units produced by Edwards were supplied to the United Kingdom Atomic Energy Authority. So we find that the initial work on EBW in France, the USA and the UK was associated with nuclear applications, and that the pioneering work was undertaken with essentially the same type of equipment employing accelerating voltages of up to 30kV.

However, before and during the period dealt with above, an entirely independent development was taking place in West Germany. This was markedly different in that it originated from a different premise. In 1948 or thereabouts, while the technology of Germany was still restricted by postwar conditions, K. H. Steigerwald was conducting research work in oscillography at very high frequencies. He was forced to develop many of his own components, forced even to use electrostatic lenses because iron and copper of the quality required for magnetic lenses were simply not available. He noted, as others had also reported earlier, that the high-intensity electron beam he needed was apt to melt or even erode away the anode of his equipment.

His observations and reports to the directorate of the Suddeutsche Laboratory, in which he worked, on the possibilities of his equipment as a heat source met with little response or encouragement. The joint owners of this laboratory, AEG and the Carl Zeiss Foundation, could see no practical outlets at that time. However, a patent application[3] was filed in January 1951 which described electron-beam equipment for drilling holes in various materials. An additional patent[4] was filed later in 1951 by the AEG patent lawyer Schneider, which not only referred to welding but also envisaged the idea of non-vacuum EBW. There is no doubt that the drilling equipment was fully capable of seam welding even at that time. Probably it had sufficient intensity to have been capable of demonstrating the new phenomenon of deep-penetration welding, although there seemed no need in Germany at the time for producing welds in vacuum. But all this had to come later as work was abandoned in 1951 when the laboratory was disbanded, and Steigerwald left Zeiss for the first time.

At this time a Mr Irving Rossi of New York, best described as a 'technological scout', came into the picture. Rossi is a central figure in the history of EBW in that he had the vision to recognize the technological potential of Steigerwald's work when he saw it, stepping in just as the work was probably about to be completely abandoned. In 1951 he obtained the world rights for 'drilling of fine holes in solid bodies, ... melting, soldering and welding in vacuum utilizing an electron beam' from the AEG-Zeiss interest and assisted Steigerwald by financing the manufacture of two machines (see Fig. 1.3) with which to follow up the potentialities of a well-focused high-voltage electron beam. These two machines, capable of the high accelerating voltage of 125kV, and of 2·5kW output were intended primarily for drilling small holes such as are required in the man-made-fibre industry. However, as drilling required only some 100W, the greater power output of the machine was intended for exploring other applications such as melting and welding. The machines were built at Hanau by Steigerwald in a section of a factory jointly owned by Siemens and Heraeus.

But even then the project developed little further. The two machines, when built, stood in water in a leaky shed until they were moved to a watchmaking factory owned by Junghausen in the Black Forest. There, Steigerwald continued his work, drilling jewelled bearings, spot welding thermo-elements, cutting, melting, soldering, and welding metals and ceramics. Left over at Hanau was an interest by Heraeus in the more general possibilities of the thermal effects of an electron beam, but we shall return to this later.

By this time Rossi realized that, if the project were ever to come to fruition, more backing would be needed. In 1954, he negotiated a contract with Zeiss who then came back fully into the picture. Work continued on a more vigorous scale and the now familiar electron-optical column of the Zeiss machines, which was to remain unchanged for ten years, was developed in 1956 as shown schematically in Fig. 1.4. Even then, the technique was only a solution in search of a problem. But the stage was set for the linking up of this very advanced equipment to the need of the nuclear industry.

This came about when Westinghouse had heard that Steigerwald had developed novel electron-beam equipment at Oberkochen. They had also heard the news from Paris of Stohr's initial work on

Fig. 1.3 Improved version of 1952 electron-beam equipment which operated at 125kV and was rated at 2·5kW

electron-beam welds in the French nuclear energy industry. But the requirements of Westinghouse were even more precise than the type of welds Stohr had produced. Their objective was to produce a series of precision welds close to each other in a Zircaloy reactor

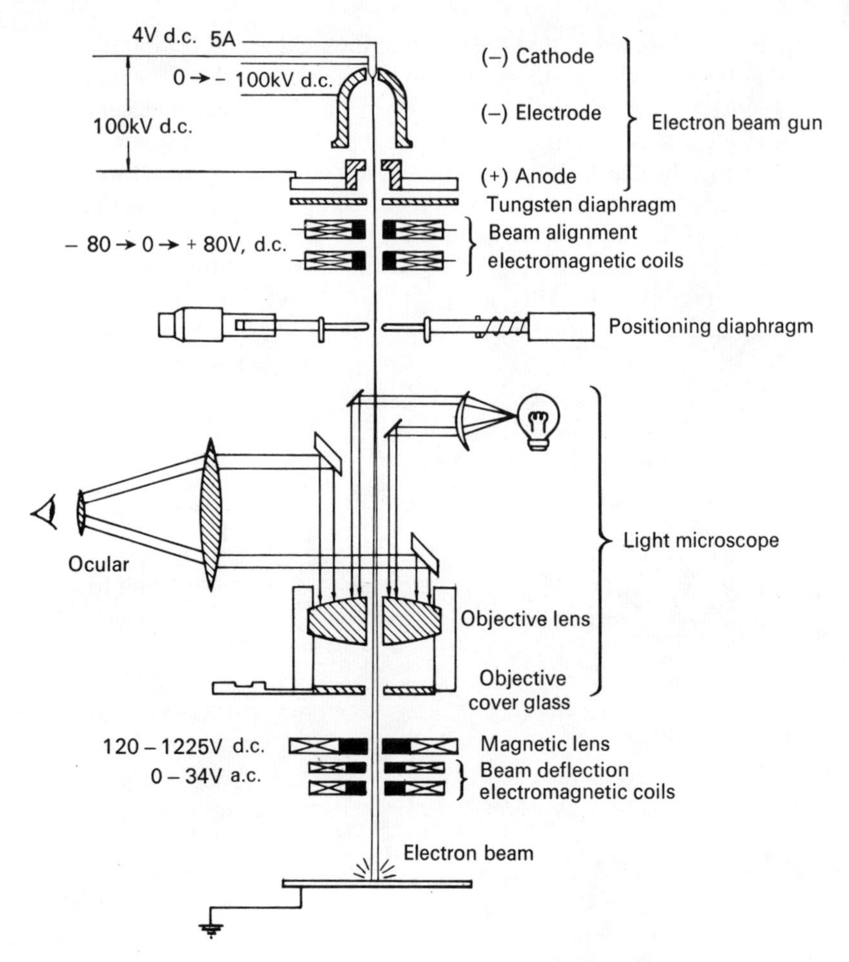

Fig. 1.4 Early Zeiss equipment as used by Burton and Matchett[5]

component. It could not have taken long for Steigerwald to demonstrate that his equipment was capable of meeting this requirement. He already had the equipment which he had developed from the electron microscope, and it was only a simple step to produce electron-beam welds and thus reproduce the welding work of Stohr. The success of the demonstration heralded the interest of the USA and paved the way to a major commercial involvement.

Welding by the electron beam was a new experience in the Steigerwald Laboratory. He observed, what had been hinted at earlier in Stohr's work, that the depth of the weld was substantially greater than the width. However, with the high power densities that Steigerwald could achieve with his equipment, which were necessary for drilling fine deep holes, and with the lower thermal conductivity of Zircaloy, the depth-to-width ratio of the weld then produced was considerably greater than had been reported by Stohr. This was the beginning of the 'deep penetration' era.

One of the two 125kV, 2·5kW machines was air-freighted to Westinghouse by 1958. In February 1959, Burton and Matchett described the equipment and its use at the Bettis Atomic Power Division.[5] Further units were quickly ordered from Zeiss. The project was off the ground and Zeiss found they had to concentrate on building machines rather than on further development work.

Around 1960, Hamilton Standard, a Division of United Aircraft, took a direct interest and formed an association with the Zeiss-Rossi-Steigerwald partnership for further development in the USA. The European interests recognized the value of the participation by Hamilton Standard and the substantial effort they would bring to bear on new applications. These were sparse in Germany, short as it was of aerospace, atomic energy, and aero engine industries.

By the autumn of 1963, Zeiss and Rossi had relinquished their interests in the enterprise to Hamilton Standard, and Steigerwald was free to start his own company to pursue the development of yet more advanced equipment for drilling and welding.

To return once more to Hanau, the interest of Heraeus in electron-beam technology, which was stimulated by the performance of the early Steigerwald machines, opened up possibilities for melting refractory metals under high vacuum. This would add usefully to the Heraeus range of vacuum arc melting furnaces. In 1952 Schumacher had, at Stuttgart Technische Hochschüle, developed and patented a system of dynamic seals through which he could allow a beam of charged particles to emerge into air. His main concern was the study of the fluorescence of gases at reduced pressures when excited by those particles. At the time, his electron gun could not deliver a sufficiently finely collimated beam, but, with the assistance of Steigerwald using his telefocus gun design, Schumacher succeeded in solving this problem. Heraeus acquired the rights to the Schumacher patent and in 1956 had built a powerful device (150kV, 2·5kW) in which a focused electron beam was capable of emerging into air. There was no thought, at that time, of using the device for welding.

The idea of in-air EBW came, perhaps, in the early sixties when Heraeus succeeded in producing welded samples at that time.

Returning now to the work of Stohr, it will be recalled that a licence was granted to Edwards High Vacuum. Another licence was also arranged with Sciaky in Paris, a private company which had been manufacturing electrical welding machines for many years. The Paris division of Sciaky is closely linked with its American counterpart in Chicago and it is by this means that EBW found another *entrée* into the United States. The following is a brief account of the Stohr–Sciaky association which ultimately resulted in the manufacture of over 200 'Stohr-type' EBW machines which, although of lower voltage, are nevertheless of high power and are being successfully used in a wide range of industries.

A business association between Dr Stohr and Mr Mario Sciaky goes back to the mid-thirties; Dr Stohr had developed a thyratron control system which was applied to resistance welding machines built by Sciaky. This business association developed into a personal friendship between the two men and it was only natural that Mr Mario Sciaky was given an early opportunity to examine the first machines constructed by Dr Stohr. The CEA requirements, now that EBW had been successfully demonstrated as a useful tool in the manufacture of reactor cans, were for plant suitable for quantity production. Sciaky designed and constructed such equipment; indeed the Stohr Laboratory contains some unique and advanced machines built by Sciaky, and this arrangement continues to be in operation.

Mr Mario Sciaky saw the industrial potential offered by this new technique and invited his brother Mr David Sciaky, who controls the American division of the company, to join him in examining Dr Stohr's work. Both brothers were equally enthusiastic and a licence agreement was quickly concluded between the CEA and Sciaky in August 1957. The first of the Sciaky production machines employing a mobile gun inside the vacuum chamber was constructed in 1959, followed by a second machine in 1960 which was installed at the Saclay plant of the CEA.

Intensive development work was carried out by Sciaky, both in Chicago and Paris, and many machines were built, all of the mobile type, placed within the vacuum chamber. The early guns were of 30kV accelerating voltage, but these were later superseded by guns operating at 60kV which substantially increased the flexibility of application of the plant, particularly with regard to the working distance, which was markedly increased by the application of the higher voltage.

In the United Kingdom, Sciaky produced their first machine in 1961, based on the combined French–American design. A number of such machines have been installed in various centres in the UK.

For convenient reference, some of the key dates in the history of EBW are shown in Chart 1.1. Some earlier dates are also included commencing with the first recorded demonstration in 1879 by Sir William Crookes[6] that an electron beam could melt metal. Sir William fused the platinum anode in a cathode-ray tube but it was nearly twenty years later when, in 1897, J. J. Thompson[7] showed that cathode rays were in fact beams of electrons. The first serious attempt to use the electron beam as a tool for melting appears to be that recorded by von Pirani in 1907.[8]

Chart 1.1 History of EBW development

1879—**Sir William Crookes:** fused a platinum anode in cathode ray[6]
1897—**J. J. Thompson:** showed that cathode rays were electron beams[7]
1907—**Marcello von Pirani:** patent on electron-beam melting[8]
1938—**M. von Ardenne:** use of electron beams as a work tool
1951—British Patent 727,460: Fine-wire spot welding by electron beam[4] (forerunner of Zeiss equipment); **Steigerwald** constructs EBW machines, primarily for drilling
1954—**J. A. Stohr:** begins work on electron-beam welding
1956—**Heraeus:** produce a non-vacuum electron gun

1957—**Stohr** reports work at Fuel Elements Conference in Paris, November

1958—**Wyman** reports work in USA;[2] **Zeiss** equipment supplied to Bettis; **Olshansky** reports work in USSR; first commercial equipment available in UK
1959—First report of welds with very high depth-to-width ratio[10] using **Steigerwald** equipment
1960—**Hamilton Standard** association with **Zeiss**
1961—Low-voltage systems developed to produce deep-penetration welds; wider diversification of EBW covering many applications
1962—Non-vacuum work commences at **Hamilton Standard**
1963—Closed-circuit TV viewing systems used; **Zeiss** relinquish EBW interests to **Hamilton Standard**; **Steigerwald** forms own company
1969—Increased sophistication, improved detail, larger equipment, and 'soft' vacuum

We have now examined in some detail the early developments which may be considered the most significant in the history of EBW and have seen how the demands of nuclear energy gave rise to the clear emergence of the technique. However, the requirements of nuclear energy alone could not justify the rapid development and widespread acceptance of EBW which has since taken place. We must therefore bring the history up to date and assess other factors which contributed to its growth.

As we have seen, the technique was initially developed as a means of fusion welding in vacuum and as such was particularly suited to joining refractory metals. Although there were few industries that dealt with such materials, it was nevertheless obvious that EBW had certain attractive features which made it potentially useful. Fine, deep welds, extremely neat in appearance, could be produced even if the need to work in a vacuum was not necessarily the main attraction. Reference was made earlier to deep penetration welding, but this was usually referred to in comparison with TIG-welding and normally implied depth-to-width ratios no greater than 2:1. Indeed, Burton and Matchett[9] in May 1959 refer to this value as a 'remarkably favourable fusion zone geometry'. The first reference and photograph of a weld section with a depth-to-width ratio of the order of 20:1 appears to be an anonymous article[10] in *Metalworking Production* of 20 November 1959, i.e., two years after Stohr's disclosure of the process. Even though this is a very important feature of the EBW process, sufficient interest in the technique had already been induced so that several manufacturers, particularly in the USA, had produced equipment for general sale. As mentioned earlier, the first commercial equipment in the UK was produced towards the end of 1958. All early units available were essentially low-voltage equipment following the original system disclosed by Stohr. Equally, most of the early units were installed at establishments concerned with nuclear energy applications although they were occasionally used to examine the electron-beam weldability of more common materials to ascertain the potentialities of the technique.

More plant manufacturers had entered the market by 1959 but it was not until 1960 that the feasibility of high depth-to-width ratios, of the order of 10:1 or better, really began to be publicized and at this time it was associated only with high-voltage systems ($\sim$100kV). However, developments of low-voltage systems continued, particularly in the area of increasing total power. About a year later, references began to appear in the technical press and elsewhere showing that guns operating at low voltages were also capable of producing welds of high depth-to-width ratio.[11,12] It was probably not until this stage had been reached that the EBW process really began to demonstrate its full potential for a wider range of applications.

Thus, it was probably not until the start of the 'sixties that the process began to enter other fields where more conventional techniques were proving inadequate. This diversification covers three main industries: aerospace, electronics, and instrumentation.

It is the last area which covers the first recorded non-nuclear production application of EBW in the United Kingdom. In January 1961, Appleby and Ireland Ltd took delivery of a 20kV, 100mA unit for use in the manufacture of aneroid capsules and similar components. In this case the electron beam was ideally suited to the production of evacuated capsules in thin-gauge materials, and to making joints inaccessibly placed between narrowly spaced parts. This work is discussed in more detail in a later chapter.

The first high power density equipment in use in the UK, a Zeiss unit, was installed by Rolls-Royce in mid-1961. Here, the initial interest in the process was the ability to obtain narrow fully penetrating welds in materials used in aero engine manufacture. It was not the intention to replace conventional processes where they were already being used successfully but to concentrate on areas which were not adequately covered, such as the welding of components in the higher Nimonic range of alloys. Although there was no requirement for very deep penetration welds, there was a need for minimum heat-affected zones with associated reduction in thermal distortion of the component. The aerospace industries were at this stage becoming one of the main non-nuclear users of the EB process and during succeeding years were responsible for much of the growth of the technique. In the USA particularly, the process was also building up a reputation in salvage work where complex, expensive, and accurately machined components, which had been rejected, could be reclaimed.[13]

The requirements of the electronics industry, particularly in the microcircuitry field, probably provided the first-ever non-nuclear application and gave a further boost to the EB process in both cutting and joining modes and contributed to its growth primarily in the USA.[14] Although cutting has been investigated in the UK there are no references to the use of EB joining in the electronics field. In the USA some sophisticated automation techniques have been examined for high-speed production of the large number of joints often involved. Figures of 1600 joints in 1in^2 (625mm^2) have been quoted.[14]

From the inception of the process, developments have been aimed at increasing both total power and power density together with the general sophistication of the equipment. Figure 1.5 shows two examples of the latest commercially available equipment. The increase in power and the use of high-voltage and low-voltage systems have already been discussed. One particular area of general improvement which may be mentioned is the development of methods of viewing the welding and aligning the beam with the weld joint. This is important because of the precise nature of an EBW operation. On early equipment this relied upon viewing through a window in the welding chamber, sometimes with the aid of a telescope. The disadvantage of this for consistently accurate work soon became apparent and alternative means had to be found. High-voltage equipment from its early days had always incorporated an optical system to allow viewing coaxially with the

Fig. 1.5 Two examples of recently produced machines: (a) Hawker Siddeley 6kW machine (*Courtesy Hawker Siddeley Dynamics, Electron Beam Division*), and (b) Sciaky 60kV/250mA machine (*Courtesy Sciaky Electric Welding Machines Ltd*)

electron beam at some magnification, in a manner similar to that used in the electron microscope. Since then, other and similar optical systems have been introduced for all types of unit. A further development is the introduction of closed-circuit television, which was probably first tried around 1963 using conventional equipment. However, more recently, miniature videcons, $\frac{1}{2}$in. diameter (12·7mm) have been developed which are capable of being fitted and operated inside the vacuum chamber.

These developments, along with larger chambers, quicker evacuation, and engineering for production may be considered as general development of the equipment. If we look for more radical advances, there is probably the non-vacuum welding technique which was first achieved by Heraeus. Hamilton Standard began work on non-vacuum EBW in mid-1962 and first reported the results nearly two years later.[15] Worth-while applications of the technique are claimed, including such possibilities as the continuous

(b)

fabrication of tubes and the welding of missile cases. Few detailed reports of any true production applications have yet been published and it remains to be seen how great an impact this variation of the technique will make.

Many of the advantages of vacuum EBW are retained, without the complications of the non-vacuum method, if welding is carried out in a 'soft' vacuum environment. This method is rapidly gaining acceptance and has a number of attractive features; it is described in greater detail later in the book. This new development may prove to be one of the more significant recent advances in EBW technology.

REFERENCES

1. Stohr, J. A. Fuel Elements Conference, Paris, November, 1957, 18–23.
2. Wyman, W. L. *Weld. J. Res. Sup.*, (USA), **37** (2), 1958, 49s–53s.
3. British Patent No. 714,613 (Application date 31 January, 1951).
4. British Patent No. 727,460 (Application date 8 September, 1951).
5. Burton, G., Jnr and Matchett, R. L. *American Machinist*, **103** (4), 23 February, 1959, 95–8.
6. Crookes, W. *Phil. Trans. Roy. Soc.*, Part I, 135; Part II, 641, 1879.
7. Thompson, J. J. *Phil. Mag.*, 1897.
8. von Pirani, M. US Patent No. 848,600 (26 March, 1907).
9. Burton, G., Jnr and Matchett, R. L. 'Electron beams—new technique for welding', *Metalworking Prod.*, **103**, 22 May, 1959, 901–4.
10. Anon. *Metalworking Prod.*, **103**, 20 November, 1959, 1879–82.
11. Anon. *Engineer*, **212** (5503), 1961, 78.
12. Stohr, J. A. 3rd Symposium on Electron-beam Technology, 1961, 102–15.
13. Meier, J. W. ASTME Welding and Metal Joining Seminar, Cleveland, Ohio, January 1963.
14. Meier, J. W. *Weld. and Metal Fab.*, **30** (7), 1962, 265–6.
15. Meier, J. W. *Weld. J.*, **42** (12), 1963, 963–7.

2. Generation and control

This chapter is confined to electron guns of the conventional form in which electrons are derived from a solid cathode. Such guns are almost universally employed on EBW machines today. More unconventional beam-forming systems making use of a cold-cathode discharge or operating at low gas pressures are feasible and have been demonstrated; indeed, they might lead to an eventual breakthrough in the design of EB guns. The field, however, is too diverse and full coverage for such systems would not be appropriate here.

The present treatment briefly reiterates some of the fundamental properties of electron guns and describes the design and performance of particular guns suitable for welding. The beam system as a whole, Fig. 2.1, rather than the electron-generating gun in isolation, is then considered in the latter part of the chapter.

The primary function of the electron-optical system of an EBW machine is to produce a controllable high-power electron beam which may be focused into a small spot on the workpiece. The design of such a system is directed towards achieving a high power density in the focal spot leading to large 'penetration'* and depth-to-width ratio of the weld. Power densities of several MW/cm^2 are usual.

There are some fundamental as well as practical limitations which prevent the attainment of an infinitely small focal spot and thus infinitely large power density. The most important of these are:

(a) the mutual repulsion between individual electrons, i.e., the space-charge effect;
(b) the existence of a small but finite initial velocity of the electrons leaving the cathode, i.e., the thermal limitation;
(c) the spherical aberration of the spot-forming electron lens.

* The term 'penetration' as used here is that now generally accepted as referring to the depth of the bore hole produced by the electrons; it is not the electron range in the material.

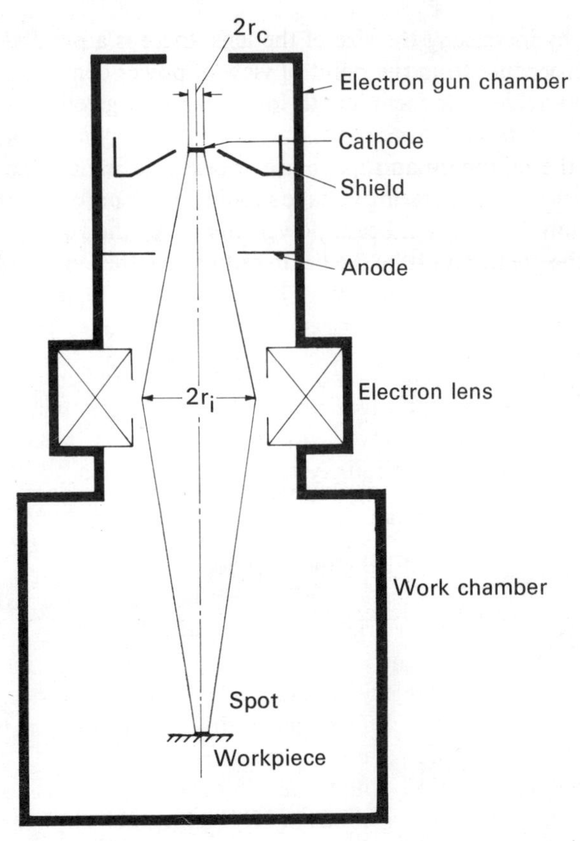

Fig. 2.1 Electron beam and main components of an EBW system

A compromise solution which does not allow any one of these spot-degrading factors to become excessive is necessary. It is particularly important that the space-charge effect should not become the predominating factor within the range of beam current of interest. If this occurs, the focal spot size increases so rapidly with beam current that the power density is reduced and further increase in power fails to produce deeper weld penetration.

It is shown later in this chapter that both the space-charge and thermal limitations are reduced by an increase of beam radius r_i at the location of the lens, and that both vary inversely as this radius. However, the spherical aberration introduced by the lens increases as the cube of this beam radius. Although this latter factor may be reduced by increasing the size of the lens, there is a practical limit to this procedure from the point of view of power demand and the minimum achievable focal length for a lens of a given dimension.

These effects and limitations, together with other parameters such as the minimum and maximum working distance, determine what particular accelerating voltage should be chosen for a particular application. For a given beam power, the lower the voltage that is chosen the greater must be the beam radius r_i in the lens, and hence the greater the lens dimensions to achieve the same spot size. This in turn limits the minimum working distance. In general, therefore, the high-voltage beam offers rather greater flexibility since the available range of throw distance for a given maximum spot size is larger.

Too great an emphasis must not be placed simply upon the achievement of minimum spot size. It is often worth accepting spot sizes somewhat greater than the smallest achievable in order to gain on other parameters, e.g., cathode life, simplicity of equipment, etc. Little is lost in penetration depth or depth-to-width ratio of the weld, as these are but weakly dependent on spot size in the practical range of 10^{-2}–10^{-1}cm dia. Figure 2.2 is a typical set of experimental results for stainless steel illustrating how the same penetration can be achieved over a wide range of spot diameter (2×10^{-2}–10^{-1}cm) by making the necessary adjustments to beam power and welding speed. The range of beam power varied between 1 and 9kW and welding speed between 0·06 and 3·3cm/s.

The application of electron beams to welding results in systems which lend themselves to sophisticated control of the welding process. A first step towards the realization of effective beam control is an accurate measurement of important process variables, such as beam power and beam focus, which are then used as control parameters in closed-loop electronic circuits. An outline of some of the control techniques employed is given in the last section of this chapter.

Electron-beam generation

Electron emitters

The cathode in an electron gun is required to provide a copious stream of electrons and to remain unaffected by the environmental conditions to which it may be subjected. In many EBW machines,

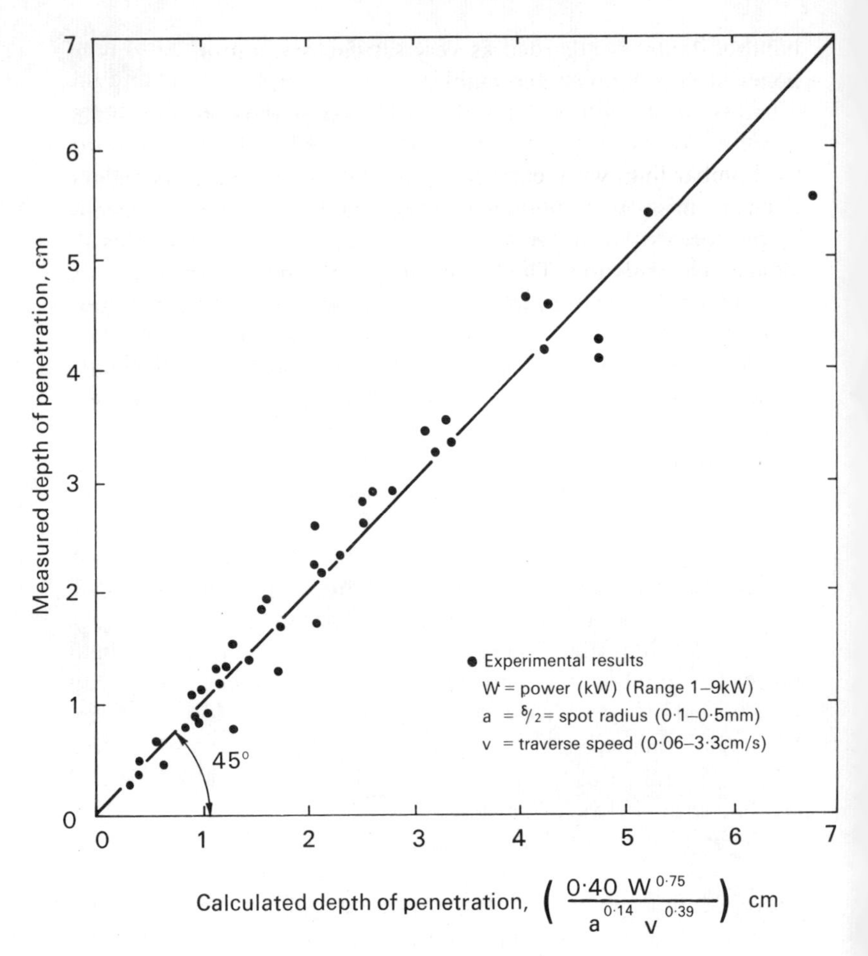

Fig. 2.2 Depth of EB weld penetration in stainless steel

where there is little isolation of the cathode region from the work chamber, these conditions may be obnoxious owing to gas and vapour released from the workpiece. The choice of cathode is then limited to that which will maintain stable emission and have an adequate lifetime under these conditions. The actual lifetime considered satisfactory depends upon the duty envisaged, time and ease of cathode replacement, reproducibility of electron-optical conditions upon replacement, and time taken to recover stable operation of the gun after a changeover. In general, a life of a few hundred hours is regarded as very satisfactory, and a few tens of hours as the minimum acceptable.

Emission of electrons from the cathode material can be brought about in several ways: by heating to a high enough temperature, by bombarding with energetic particles to produce secondary electron emission, by illuminating with light rays to yield electrons by photoemission, or by applying a large electric stress so as to induce field emission. The first of these—thermionic emission—is the only one so far of practical interest in the case of electron guns applied to EBW.

As the temperature is increased in a metal cathode, e.g., tungsten, tantalum, molybdenum, some of the free conduction electrons within the metal acquire enough energy to overcome the potential barrier at the surface. The energy in electron volts that must be acquired by an electron before it can escape is called the 'work function' of the metal.

The law relating emission current to temperature was first proposed by Richardson[1] and later modified by Dushman:[2]

$$J = KT^2 \exp - (eE_w/kT) \quad (2.1)$$

where

J = current density, A/cm^2
K = 120·4 A/cm^2 deg^2 (The exact value varies, depending upon the cathode material and its condition.)
T = absolute temperature, °K
e = electronic charge, $1{\cdot}60 \times 10^{-19}$C
E_w = work function of the metal, V
k = Boltzmann's constant, $1{\cdot}380 \times 10^{-23}$J/°C

This current density may be extracted from the cathode by subjecting the cathode face to a suitable electric field. In the simple two-electrode gun, or diode, this is achieved by maintaining a second electrode, the anode, at a positive potential with respect to the cathode (see Fig. 2.1). If this electric field is large enough, all those electrons which escape from the potential barrier of the cathode will arrive at the anode and can be made to pass through a hole in it to form the useful beam in the field-free space beyond. Operation of the cathode in this mode where all the thermally emitted electrons are removed from the face of the cathode is referred to as 'temperature-limited' operation.

It is more usual to operate the cathode under space-charge limited conditions. In this case, an excess of electrons exists at the cathode by choice of a suitably high operating temperature. The presence of the electrons between the electrodes reduces the potential in the inter-electrode space and, with increasing beam current, the potential gradient at the cathode approaches zero or even becomes slightly negative. This imposes a limitation on the current, depending upon the electrode structure and potentials, as described in the section *Beam current control and modulation* later in this chapter.

It follows from equation (2.1) that a low work function is a desirable attribute of the cathode material. Unfortunately the better electron emitters, including the well-known oxide materials, are sensitive to gas 'poisoning' and ion bombardment. Electron-beam welding guns usually employ a pure metal cathode. The reason for this is the purely practical one of maintaining stable and repeatable emission under the indifferent vacuum conditions, the presence of metal vapour, and the positive ion bombardment encountered in EBW machines. In nearly all welding guns, whether diode or triode, cathodes are either directly heated strips, spirals, or hairpins of foil or wire, or they are in the form of a button or rod heated indirectly. In the latter instance, an auxiliary electron source is used to supply the heating power.

The temperature and choice of cathode material is a compromise between several conflicting requirements, but, in practice, the gun designer is limited to the use of tungsten and tantalum, though a case may also be made at times for other refractory metals such as rhenium.

Life of pure metal emitter

The section *Effect of thermal velocities* (p. 49) shows that, to reduce the thermal velocity limitation of spot size at the target, it is necessary to work with a small cathode emission radius r_c (Fig. 2.1). Thus it is beneficial to decrease either the geometrical scale of the gun or the emission area from the cathode, while ensuring that the anode voltage, and hence the field in the cathode region, remains adequate to draw off the emission. Maximum intrinsic emission from the cathode is required and consequently the cathode temperature should be as high as possible; the temperature limit will be set by the failure of the cathode caused either by evaporation or by chemical erosion. The higher the cathode temperature the shorter will be its useful life. Considerable data are available on the evaporation rate of metals *in vacuo*. Dushman[3] has shown that the evaporation rate measured in g/cm^2 s is given in terms of the vapour pressure p (torr) by:

$$\omega = 5{\cdot}83 \times 10^{-2} p \sqrt{\frac{M}{T}} \tag{2.2}$$

where M is the molecular weight of the material.

The vapour pressure p for the elements has been compiled by Honig *et al.*,[4] and the evaporation rates as a function of temperature for several refractory metals have been calculated from their data. The data for tantalum, tungsten, and rhenium are presented in Fig. 2.3 with the ordinate scales in cm/s representing the disappearance rate obtained by dividing ω by the density ρ.

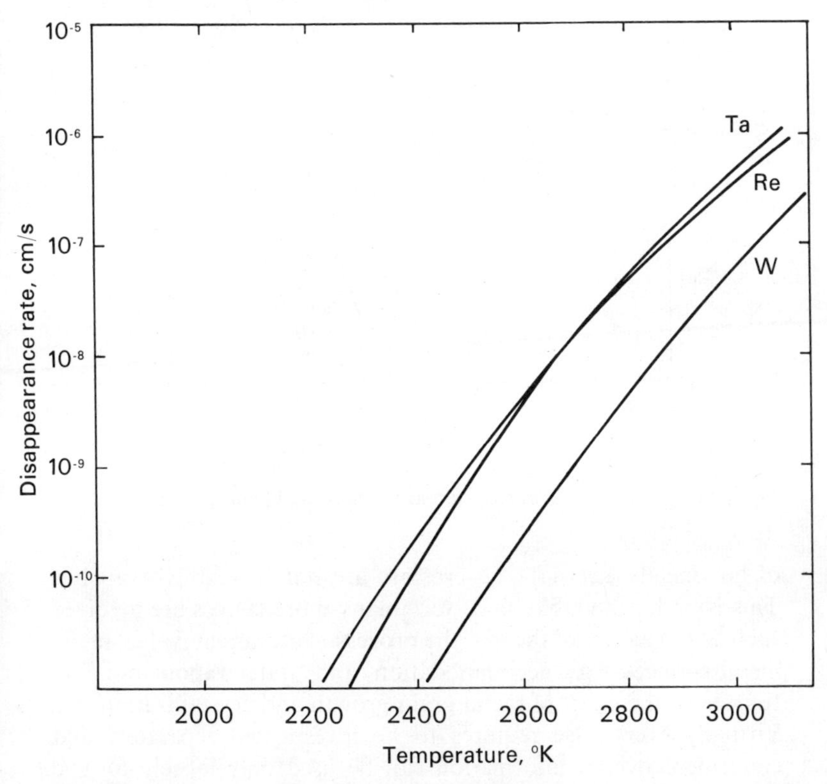

Fig. 2.3 Evaporation rates of three electron emitters

The intrinsic electron emission, i.e., the maximum current density that can be drawn at a given temperature, has already been determined for nearly all eligible metals; the curves for Ta, W, and Re are reproduced in Fig. 2.4. Data on the chemical and gas attack

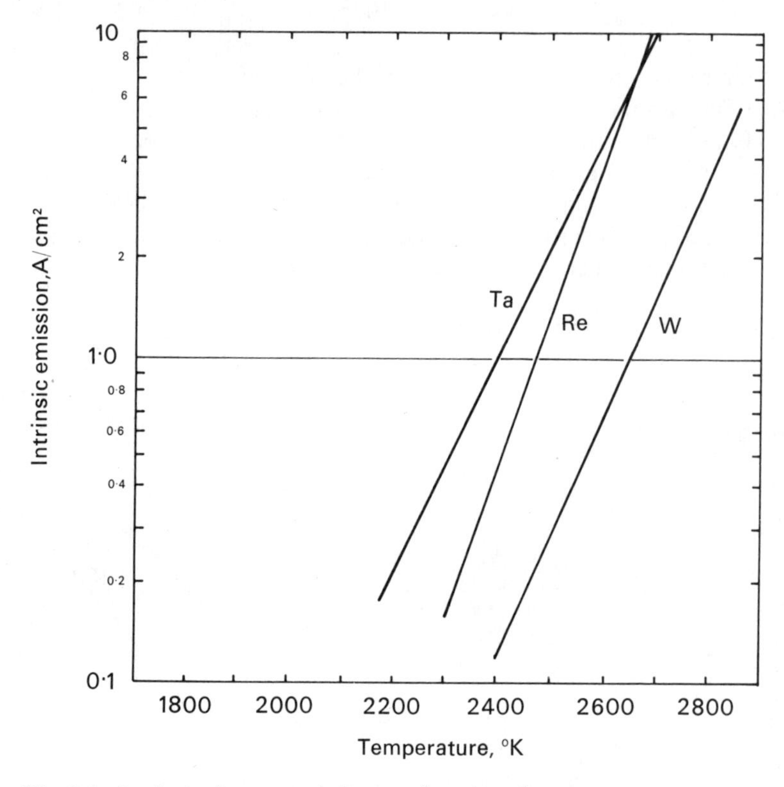

Fig. 2.4 Intrinsic electron emission as a function of temperature

of hot metals, i.e., data on erosion, are not so readily available. This is understandable, since very many more factors are involved such as the nature of the gas, the process of atom removal from the metal surface—e.g., 'accommodation' and 'water vapour cycle' in tungsten—the effect of metal grain growth and crystallization, etc. Virtually every case requires to be investigated separately and convenient overall information can be used only loosely to give general guidance.

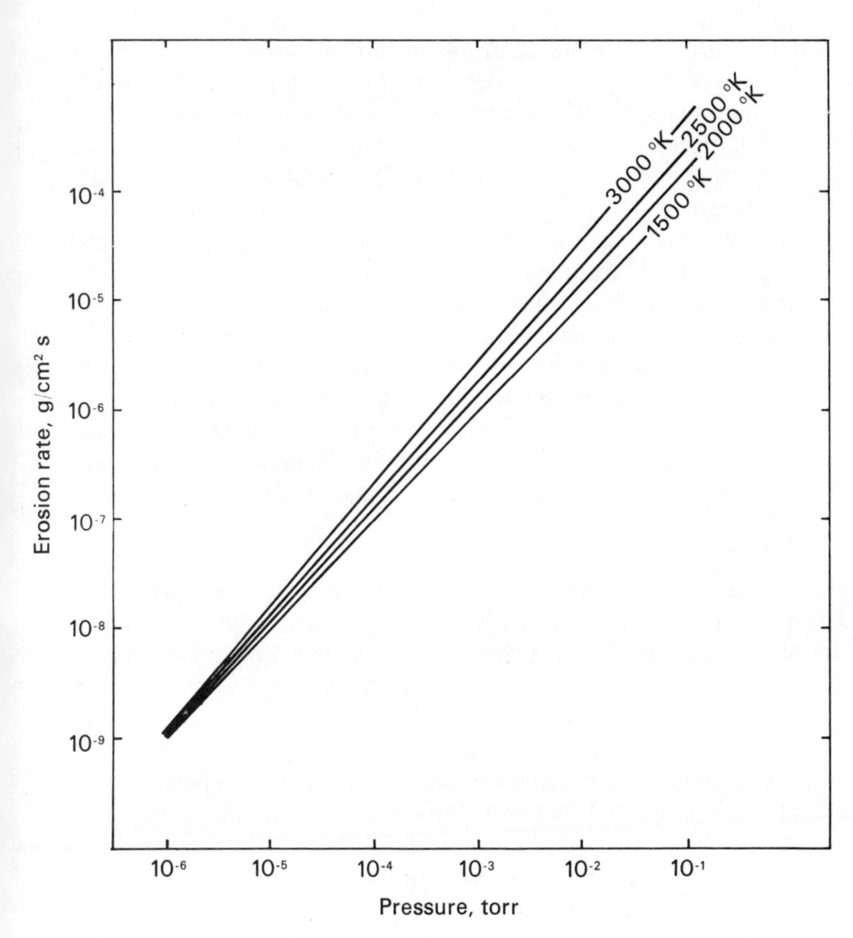

Fig. 2.5 Erosion rate in water vapour

Nevertheless, very useful general data are those abstracted from Bloomer[5] and Heavens[6] who have measured erosion rates of W and Ta in the presence of water vapour and air at several pressures and temperatures. The mechanism of metal removal as stated is not immediately obvious, but the practical results which relate to the kind of vacuum conditions encountered are of value and, furthermore, the data of these two authors are in very good agreement. The abstracted results are given in Fig. 2.5.

The following example will illustrate the use of Figs. 2.3 to 2.5 in estimating the life of a cathode for a given intrinsic emission.

Life is usually terminated in a hairpin filament when its diameter is reduced by about five per cent. For example, for a 0·075cm diameter filament this corresponds to a surface loss of some 0·002cm. For tungsten, and an emission current density of 5A/cm^2, the temperature of the cathode must be 2830°K at which temperature the evaporation rate is $8{\cdot}0 \times 10^{-9}$cm/s (Fig. 2.3); if a pressure of, say, 10^{-5}torr is assumed, the erosion rate is some 8×10^{-10}cm/s (Fig. 2.5), giving a total disappearance rate of $8{\cdot}8 \times 10^{-9}$cm/s and hence a life of about sixty hours. If the filament were made from tantalum, the corresponding temperature would become 2600°K, the evaporation rate $5{\cdot}0 \times 10^{-9}$cm/s, the erosion rate 9×10^{-10}cm/s, and the life ninety hours, indicating that there is some advantage (in life) in using Ta in preference to W. Table 2.1 compares some life figures calculated in this way for a 0·075cm diameter filament made from Ta, W, and Re, at several values of intrinsic emission.

Electron acceleration and beam profile

Though the properties of electron guns have been well and often described in the literature, a brief recapitulation of the essential basic considerations is given to assist the user in appreciating the relevance of certain design features to his specific range of application. One starting point for electron gun design is to consider electron flow between electrodes of simple geometry.

Electron flow between the electrodes of an infinite-plane, parallel diode, satisfies the Child–Langmuir space-charge law,[7–9] where the current density in A/cm^2 is given by

$$J = \frac{2{\cdot}335 \times 10^{-6} V^{\frac{3}{2}}}{x^2} \tag{2.3}$$

where

V = potential between electrodes in V
x = distance between electrodes in cm.

This expression can be simply derived from Poisson's equation, the energy/velocity relation, and the relation between energy, current density, and velocity, and gives the maximum or space-charge limited current density that may be drawn from a cathode in a simple diode. The current density is seen to vary as the three-halves power of the voltage.

Electron flow between electrodes arranged in the form of concentric cylinders (with electron flow taking place from inner to

Table 2.1 Life of 0·075cm dia. filament (5 per cent reduction in dia.)

Specific emission (A/cm²)	Ta				W				Re	
	Evaporation rate (cm/s)	Approximate erosion rate at 10^{-5}torr (cm/s)	Evaporation life only (h)	Evaporation plus erosion life (h)	Evaporation rate (cm/s)	Approximate erosion rate at 10^{-5}torr (cm/s)	Evaporation life only (h)	Evaporation plus erosion life (h)	Evaporation rate (cm/s)	Evaporation life only (h)
0·5	3×10^{-11}	8×10^{-10}	19000	670	$1{\cdot}4 \times 10^{-10}$	8×10^{-10}	4000	590	$2{\cdot}5 \times 10^{-10}$	2200
1	$1{\cdot}5 \times 10^{-10}$	$8{\cdot}4 \times 10^{-10}$	3600	560	$4{\cdot}6 \times 10^{-10}$	$8{\cdot}1 \times 10^{-10}$	1200	440	7×10^{-10}	800
2	7×10^{-10}	$8{\cdot}8 \times 10^{-10}$	800	350	$1{\cdot}6 \times 10^{-9}$	$8{\cdot}2 \times 10^{-10}$	350	230	$1{\cdot}9 \times 10^{-9}$	290
5	5×10^{-9}	$9{\cdot}0 \times 10^{-10}$	111	94	8×10^{-9}	$8{\cdot}5 \times 10^{-10}$	70	63	$5{\cdot}5 \times 10^{-9}$	100
10	$1{\cdot}6 \times 10^{-8}$	$9{\cdot}6 \times 10^{-10}$	35	33	$2{\cdot}6 \times 10^{-8}$	$9{\cdot}0 \times 10^{-10}$	21	20	$1{\cdot}1 \times 10^{-8}$	50

outer cylinder or *vice versa*) may also be treated analytically,[10] though the solution giving the current density J in terms of voltage and geometry is no longer a simple analytical function as for the plane parallel diode.

For ratios of anode radius to cathode radius R_A/R_C exceeding about 10 the expression for the space-charge limited current per unit of axial length l becomes:

$$\frac{I}{l} \simeq \frac{2\pi \times 2{\cdot}335 \times 10^{-6} V^{\frac{3}{2}}}{r} \text{A/cm} \tag{2.4}$$

where r is the radial separation. This expression is analogous to that given in eq. (2.3).

For values of $R_A/R_C < 10$, the current density is modified by a factor dependent upon the ratio R_A/R_C.

Of more immediate interest in connection with gun design is the third and only other geometrical arrangement which is at all amenable to a comparatively simple mathematical treatment, namely the case of electron flow between concentric spherical electrodes. Here, the space-charge limited current is given by

$$I = \frac{29{\cdot}34 V^{\frac{3}{2}}}{\alpha^2} \times 10^{-6} \text{A} \tag{2.5}$$

where

$$\alpha = \mu - 0{\cdot}3\mu^2 + 0{\cdot}075\mu^3 - 0{\cdot}00143\mu^4 + 0{\cdot}00216\mu^5 \text{ and}$$
$$\mu = \log_e(R_A/R_C)$$

(R_A may be greater or smaller than R_C.)

Equations (2.3–2.5) all show the maximum (space-charge limited) current that can be drawn to vary as (voltage)$^{\frac{3}{2}}$. This is the well-known three-halves-power law and applies to diode guns of any geometrical arrangement.

Because of the rectilinear flow of electrons, the cases quoted may form the starting point for the design of practical electron guns. The simplest and best known is that designed by Pierce[11] giving a beam of rectangular cross-section. It is based on the plane parallel diode case yielding the maximum current density of eq. (2.3). Pierce has simply shown that a beam of rectangular cross-section (infinite in the x direction, finite in the y direction) will obey eq. (2.3) just as a beam of lateral infinite extent, provided the gradient and potential distributions at the edge of the finite beam are unaffected by making

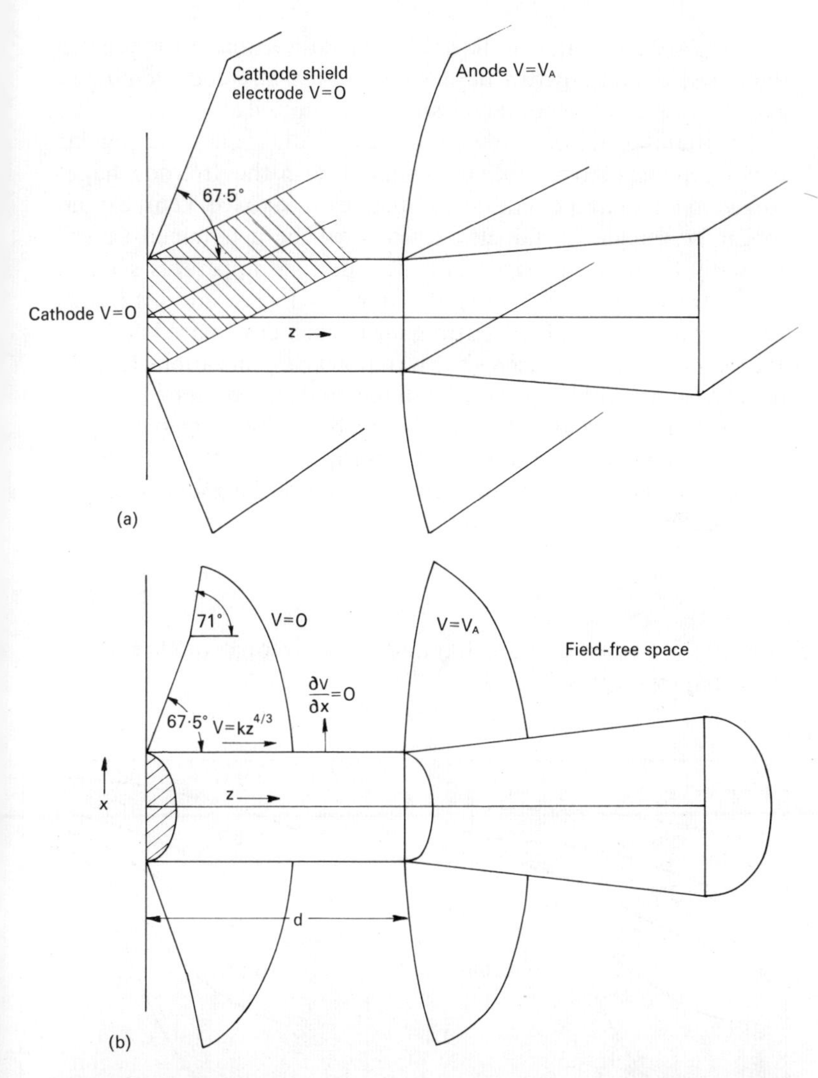

Fig. 2.6 Pierce unipotential systems: (a) rectangular semi-infinite beam, and (b) cylindrical beam

the system finite. That is, the gradient $\partial V/\partial y$ at the edge of the beam must be zero and the potential distribution along the beam edge must obey the relation

$$V = kz^{\frac{4}{3}} \tag{2.6}$$

(where z is the distance in the longitudinal direction). Pierce shows that these conditions can be derived analytically and established exactly by a set of electrodes as shaped in Fig. 2.6a.

To attain rectilinear flow of circular rather than rectangular cross-section, i.e., to produce a cylindrical beam, the electrode shapes to give the required conditions at the beam edge [eq. (2.6)] can no longer be calculated, but nevertheless can be determined experimentally by trial and error in a field-plotting apparatus such as an electrolytic tank. This was also done by Pierce, and his results are shown in Fig. 2.6b. Electron guns of such design are known as Pierce guns and have been shown in practical realizations to yield beam currents within 95 per cent of the predicted values.

There are two additional effects which take place in such guns. At the anode, where the beam enters field-free space, the beam becomes divergent owing to the lens action of the anode aperture. The focal length of this lens is:

$$f \simeq \frac{4V}{E} \simeq 4d \tag{2.7}$$

where V is the anode potential and E the average anode/cathode potential gradient $= V/d$.

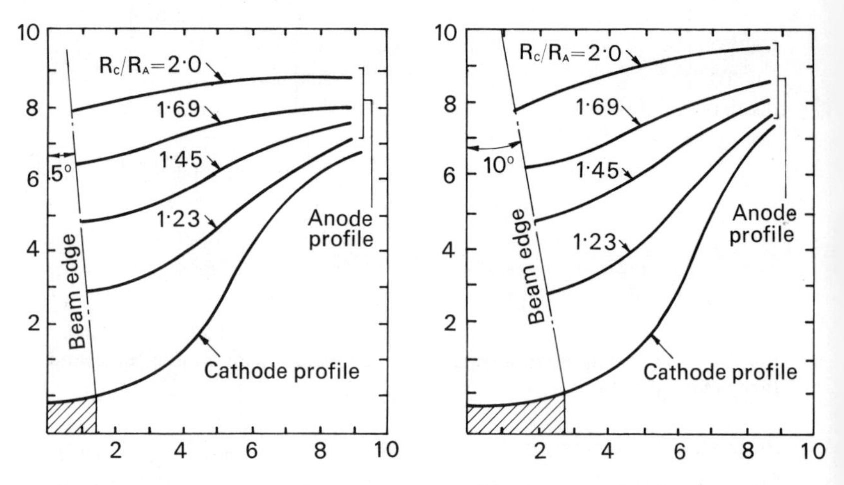

Fig. 2.7 Beam convergence in unipotential system: (a) conical beam 5° convergence, and (b) conical beam 10° convergence. Arbitary units (*after Spangenberg*)

Secondly, in the field-free region beyond the anode, spreading in the beam can take place because of the mutual or space-charge repulsion of the electrons in the beam. The magnitude of this is considered later in the section *Space-charge spreading at the target.* Both of these effects may be undesirable for certain applications and consequently it is of interest to consider the establishment of beams already possessing an initial compensating convergence, as, for example, that provided by conical sections of the concentric sphere arrangement with the anode as the inner sphere. The space-charge limited current for flow between concentric spheres is given in eq. (2.5). For a conical section of semi-angle θ (Figs. 2.7, 2.10), the current is a proportion of this, and the voltage distribution along the beam edge becomes:

$$V = \frac{1051 I^{\frac{2}{3}} \alpha^{\frac{4}{3}}}{(\sin \theta/2)^{\frac{4}{3}}} \tag{2.8}$$

[where α is the parameter incorporating distance along the beam, eq. (2.5)].

In order to achieve Pierce-type flow, the potential distribution as given by eq. (2.8) along the edge of the beam must be established, together with the condition $\delta V/\delta\theta = 0$. It is seen from eq. (2.8) that this distribution depends on the cone semi-angle θ and hence a different set of electrode shapes is required for every angle θ. Again, the required geometries can be established experimentally in an electrolytic tank by trial and error and this has been done by Spangenberg *et al.*[12] for several angles. Two geometries, for 5° and 10° convergence, are reproduced in Fig. 2.7, where several different anode locations, and hence values of R_C/R_A, are shown. Each of these shapes will impress the required potential distribution along the beam edge leading to Pierce-type flow. In designing guns on this basis, it is obviously undesirable to choose too low a value of R_C/R_A since this will lead to a departure from the desired potential distribution in the region of the anode because of the presence of the anode aperture, and may, furthermore, lead to undesirable high fields for a given anode voltage.

It is of considerable interest to assess the flow of electrons in this type of gun, especially the effect of the space-charge spreading, as well as beam divergence introduced by the anode lens. The latter can be derived readily from eq. (2.7) and a plot of b, the distance

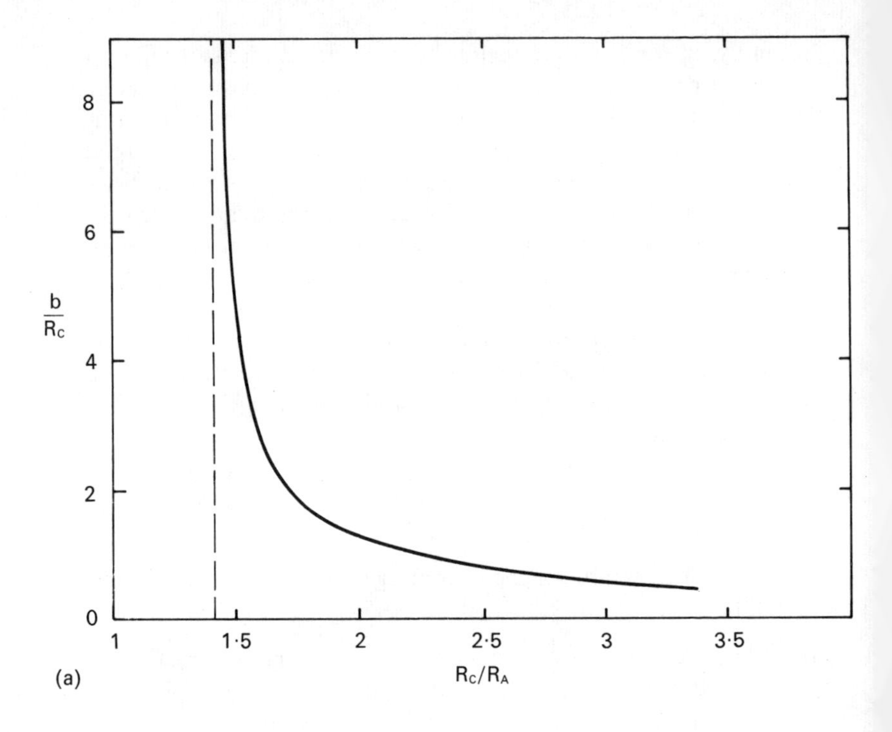

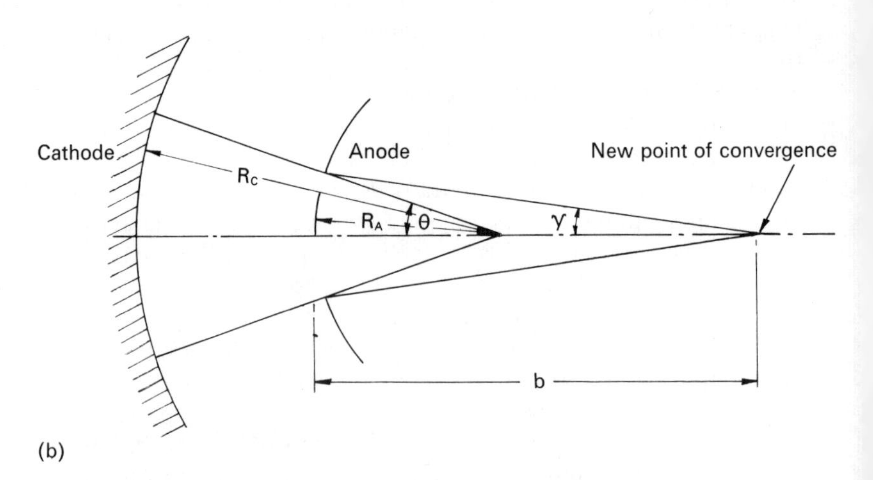

Fig. 2.8 Position of point of convergence as a function of electrode geometry (*after Spangenberg*)

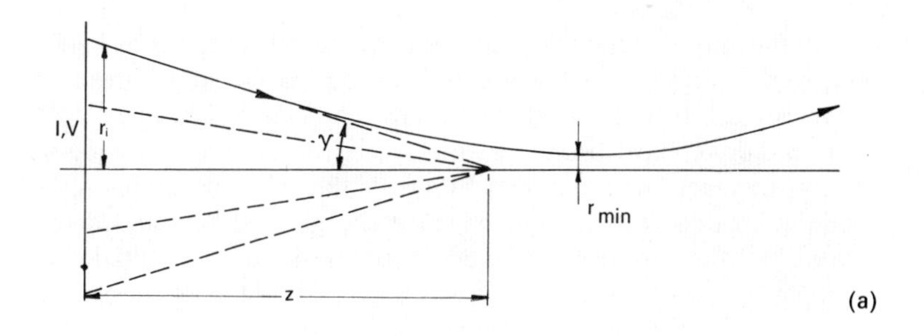

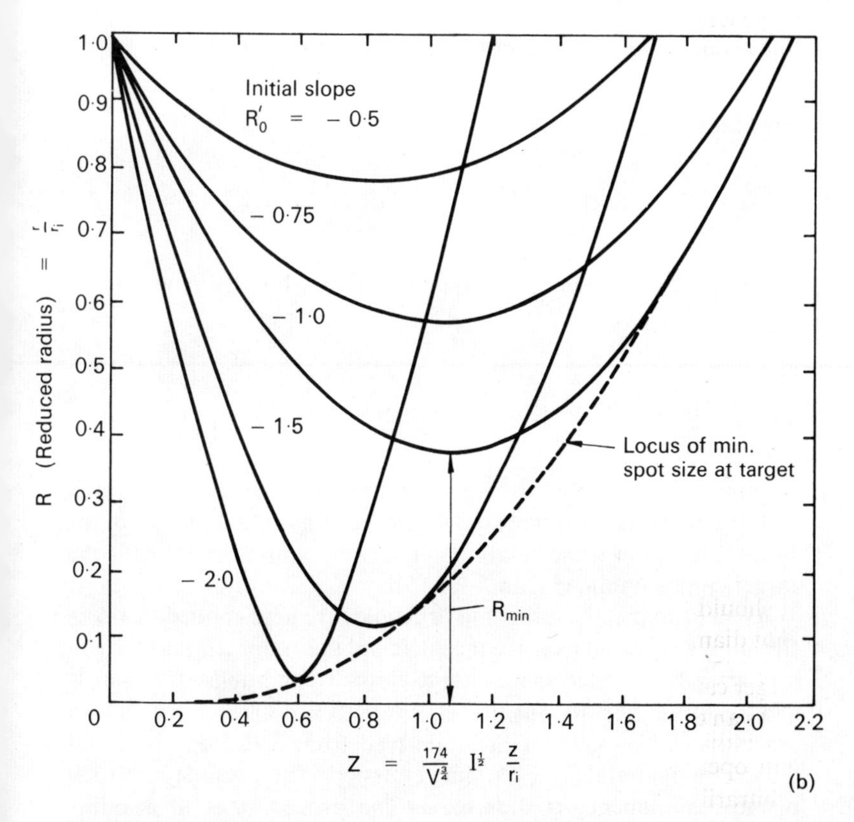

Fig. 2.9 Effect of initial convergence γ on minimum radius r_{min} (*after Pierce*)

from the anode to the new point of convergence, in terms of R_C/R_A is given in Fig. 2.8a. The magnitude of the space-charge spreading can also readily be assessed as was shown by Pierce and others.[13–15] An initially conical beam, in which the electrons are travelling in field-free space and are directed towards the axis, will attain only a minimum diameter limited by space charge repulsion and determined by the current and voltage of the beam and the initial convergence, γ (see Fig. 2.9a). A useful universal plot of beam profiles (for rotationally symmetric beams) is reproduced in Fig. 2.9b. The scales are given in 'reduced' R and Z

where $R = \dfrac{r}{r_i}$ (r_i = initial beam radius, z defined in Fig. 2.9a)

and $Z = 174 \dfrac{I^{\frac{1}{2}}}{V^{\frac{3}{4}}} \dfrac{z}{r_i}$ (I amps, V volts, r, z cms.)

The initial slopes are

$$R'_0 = \left(\frac{\partial r}{\partial z}\right)_0 \cdot \frac{V^{\frac{3}{4}}}{174 I^{\frac{1}{2}}} = \frac{V^{\frac{3}{4}}}{174 I^{\frac{1}{2}}} \tan \gamma$$

and it has been shown[13] that the minimum beam diameter is:

$$R_{min} = \exp - (R'_0)^2$$

$$\text{or} \quad r_{min} = R_{min} \times r_i = r_i \exp - \left\{\left(\frac{\partial r}{\partial z}\right)_0 \cdot \frac{V^{\frac{3}{4}}}{I^{\frac{1}{2}}}\right\}^2$$

It should be noted that for a given target distance z, the minimum spot diameter at the target ($2r_s$) is not the minimum diameter of the beam ($2r_{min}$), but somewhat greater. The minimum spot size at the target can be obtained from Fig. 2.9b.

As an example, the foregoing principles are here applied to assess the initial spreading in a concentric sphere unipotential electron gun operating under space-charge limited conditions. If a gun is arbitrarily chosen to operate at, say, 10kW, 30kV, requiring an emission of 0·33A, this can be derived from a cathode of about 0·65cm diameter at $\rho_c = 1\text{A/cm}^2$ emission. The cathode size—the prime requirement—is thus fixed. The procedure is to examine conditions as the angle θ is varied and since the current is also fixed

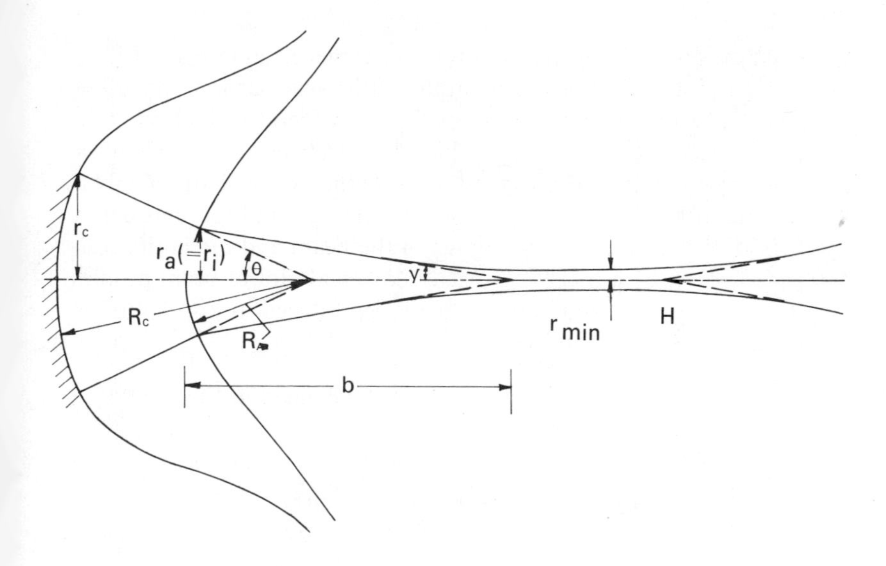

Fig. 2.10 Electrode geometry and beam profile in diode gun

this leads to a fixed ratio of R_C/R_A [eq. (2.5)] for every θ, maintaining space-charge limited emission. The initial slope γ in the field-free region beyond the anode can now be determined (Fig. 2.8b) and subsequently the beam minimum diameter. The shapes of the electrodes (Fig. 2.10) to maintain the required conditions at the beam edge vary for each θ as well as for each R_C/R_A; the correct profiles must be chosen from Fig. 2.7 for the final design. The procedure is detailed in Table 2.2. It is noted that the beam minimum (r_{min}) resulting from space-charge spreading becomes small as the angle θ is increased. At greater angles, r_{min} does not vanish, since the simple paraxial theory underlying its derivation no longer applies. Moreover, the effects due to thermal velocities now become predominant at the beam minimum.

When an image or spot is subsequently formed at the target with a projecting lens, the effective source, however, is not an image of the extended area formed by the beam minimum r_{min}. In the absence of thermal spreading, all electrons effectively emanate from a point source at H (Fig. 2.10). A point image cannot be formed at the target since space-charge spreading will occur, again giving rise to a finite spot diameter.

Table 2.2 (see Fig. 2.10) Unipotential gun $V = 30\text{kV}$; $I = 0{\cdot}33\text{A}$; $r_c = \text{constant} = 0{\cdot}32\text{cm}$

θ	3° 30′	5°	5° 8′	6°	10°	20°	30°
$\dfrac{\alpha^2 = 0{\cdot}93(V_{kv})^{3/4}\sin^2(\theta/2)}{I\ \text{(amps)}}$	0·434	0·870	0·925	1·261	3·510	13·9	30·9
R_C/R_A (eq. 2.5)	1·73	2·09	2·13	2·35	3·50	6·50	10·0
$R_C = (r_c/\sin\theta)$(cm)	5·25	3·65	3·57	3·06	1·85	0·96	0·64
R_A (cm)	3·03	1·76	1·68	1·30	0·53	0·143	0·064
b (cm) (Fig. 2.8a)	12	4	3·74	2·60	0·92	0·18	0·08
$\gamma = \left(\dfrac{\partial r}{\partial z}\right) = (R_A/b)\sin\theta$ (rad)	0·0154	0·0384	0·0401	0·0522	0·1010	0·2740	0·4120
$r_{min} = r_a \times \exp - \left[\left(\dfrac{\partial r}{\partial z}\right)\dfrac{V^{3/4}}{174 I^{1/2}}\right]^2$ (microns)	1630	690	660	330	48	—	—

Beam current control and modulation

A most useful practical modification of a diode-formed beam is the application of a negative potential to the cathode shield electrode with respect to the cathode (Fig. 2.1). This suppresses small 'edge' currents at the cathode and allows current control down to zero beam, i.e., beam 'cut-off'. In doing so, the initially unipotential gun, whether yielding a cylindrical or conical beam, now no longer operates on the Pierce principle, but may be described as a Pierce-type, near-Pierce, or triode gun. The action of increasing negative bias, illustrated schematically in connection with a cylindrical beam system in Fig. 2.11, is to move the zero equipotential over the cathode towards its centre,[16] thereby reducing or restricting the emission area and hence the beam current. An exact mathematical analysis to determine the new beam profile is virtually impossible, though fairly rough but workable approximations can be made. The important gun parameters in estimating the final attainable spot size in a complete beam system are the radius of the cathode emitting area (r_e) and the radius of the beam in the focusing lens (r_l) which depend directly on the final divergence angle of the beam from the gun. The latter is readily obtained from direct measurements. The value of r_e, the effective cathode radius (Fig. 2.11), required for an estimate of the thermally limited spot size, is given by:

$$\frac{r_e}{r_c} \simeq \sqrt{\frac{I}{I_0}} \qquad (2.9)$$

where I and I_0 are respectively the beam currents from the gun for zero potential on the shield electrode (the conventional Pierce unipotential arrangement), and with negative bias applied to the cathode shield electrode.

The application of bias produces opposing effects on the beam. The restriction of the beam current reduces the space-charge spreading in the gun, so that the beam becomes more convergent in the cathode/anode region. The location of the zero equipotential—which is regarded as the shape of the new cathode shield—may be such as to impose a greater or smaller radial field at the beam edge in this region, and consequently opposes or assists this initial convergence. In some gun designs, a balance between these two effects is observed, and the beam angle is seen to pass through a maximum as the bias potential on the shield electrode is altered

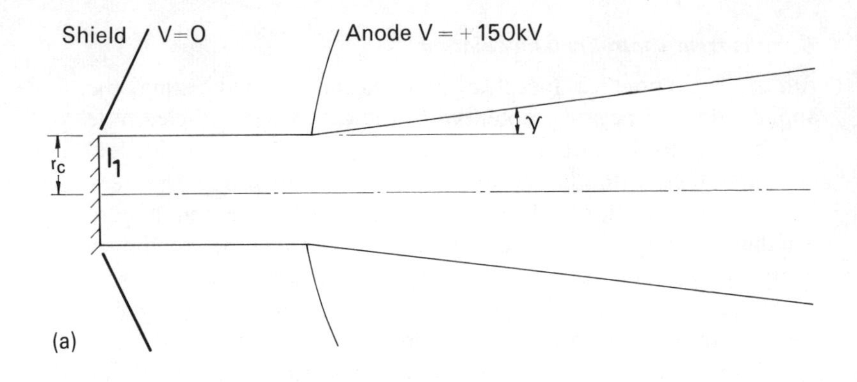

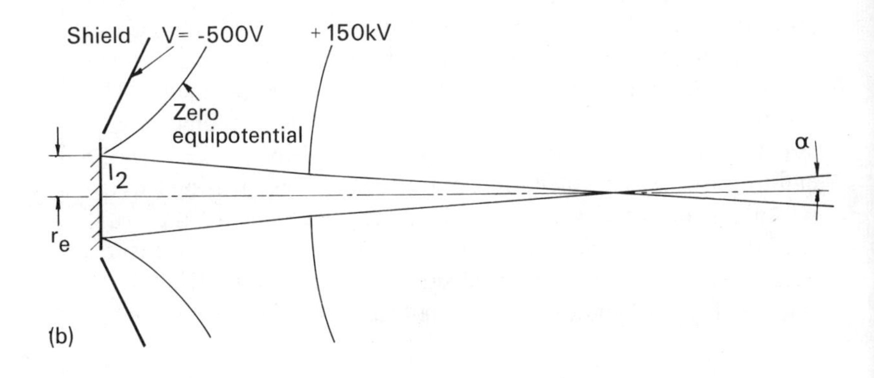

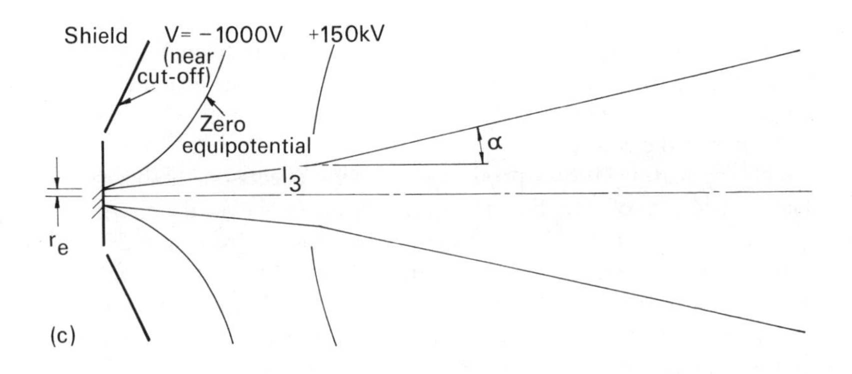

Fig. 2.11 Effect of negative bias voltage on the effective cathode radius, r_e (schematic): (a) cylindrical (Pierce), (b) triode (Pierce type), and (c) triode (Pierce type) near cut-off

from zero to cut-off. This is illustrated in Fig. 2.12 for a given arbitrary set of conditions.

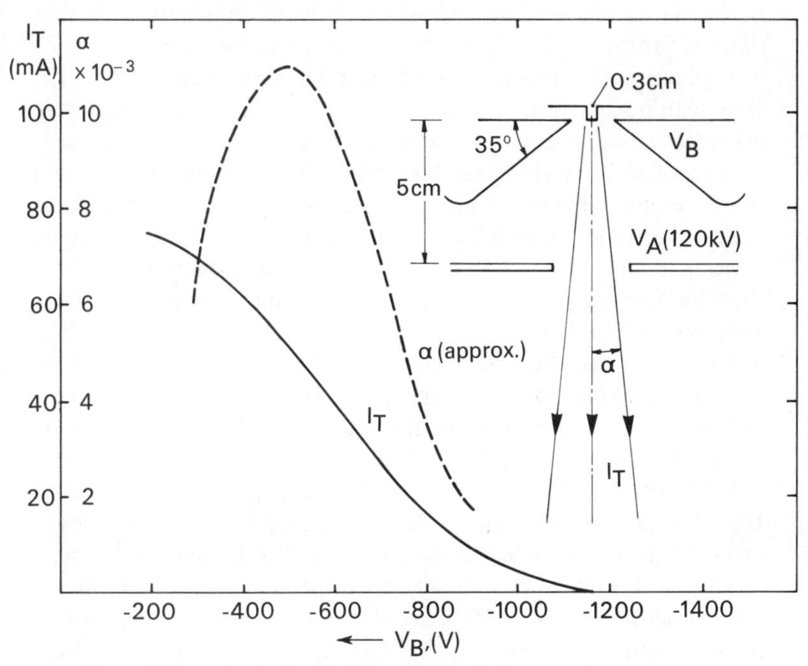

Fig. 2.12 Typical bias characteristic

In practice, by designing the electrode structure for the maximum current condition, the departure of beam profile at smaller currents, and greater bias potential, is quite acceptable.

There are many practical designs of triode gun having a variety of cathode shapes, including hairpin, and yielding a variety of relations between bias potential, beam current, and beam angle, and claims of advantage of one design as compared with others are made by various workers. As will be shown later, the important parameters, apart from beam voltage and current, when aiming to attain high spot power density are the cathode emission radius (r_e) and the radius of the beam (r_i) in the main focusing lens.

Mechanical design of welding guns

DIRECTLY HEATED CATHODES. In practice, the gun designer is generally restricted to the use of pure metal cathodes which may be either directly or indirectly heated. The former, usually hairpin filaments, are somewhat simpler to manufacture and present less of a problem in the matter of excess heat removal. There are, however, a number of disadvantages: short cathode life,* large and asymmetrical spot at high currents, and a tendency to wander mechanically off-centre, thus leading to electron spot drift. Nevertheless, such guns give acceptable performance and have been used ever since the early designs of Steigerwald in the late 'fifties. Basically, this type of gun is a development from the electron microscope gun[16–20] which also operates around 50–100kV on the triode principle though at much lower perveance ($P = I/V^{\frac{3}{2}}$), cathode size, and power. Apart from that, the design problems are the same: the gun must provide vacuum and electrical insulation to hold off cathode and bias voltage both outside and inside the vacuum, and allow the heat from the cathode to be removed. The latter is not a very severe problem in electron-microscope guns which usually employ tungsten filaments of some 125μ diameter and require only about 5–10W to heat to 2800°K. The heat removal requires more attention when the filament diameter is increased some three to five times in order to give the higher emission currents required for welding.

The biasing shield has already been the subject of comment; its chief function lies in controlling the initial beam angle from the cathode and in establishing the bias/beam current characteristic. In some designs, the lens action of the field in the cathode region can be made to converge the beam; such guns are classified as telefocus guns. The ultimate properties of the final spot at the target are, however, not critically affected by such a lens action; the vital parameters are still the radius of the beam in the lens (r_i) (Fig. 2.13), the cathode radius (r_c or r_e), and the beam current and voltage. A conventional electron-microscope gun is shown schematically in Fig. 2.14a, together with field shape and beam profiles prevailing at the filament tip as the bias potential is varied (Fig. 2.14b). Emission can take place only in the region of positive field and the effective cathode emission area is reduced to zero as the negative bias is taken to the cut-off value, exactly as in the welding gun illustrated in Fig. 2.11. A further design feature is the movable anode lens used in microscopes[21] and now incorporated in some EBW machines. The anode aperture diverging-lens effect is utilized and the beam

* Some information on life, which is limited by cathode evaporation and erosion by gas attack, is given by Bloomer[5] and Heavens.[6]

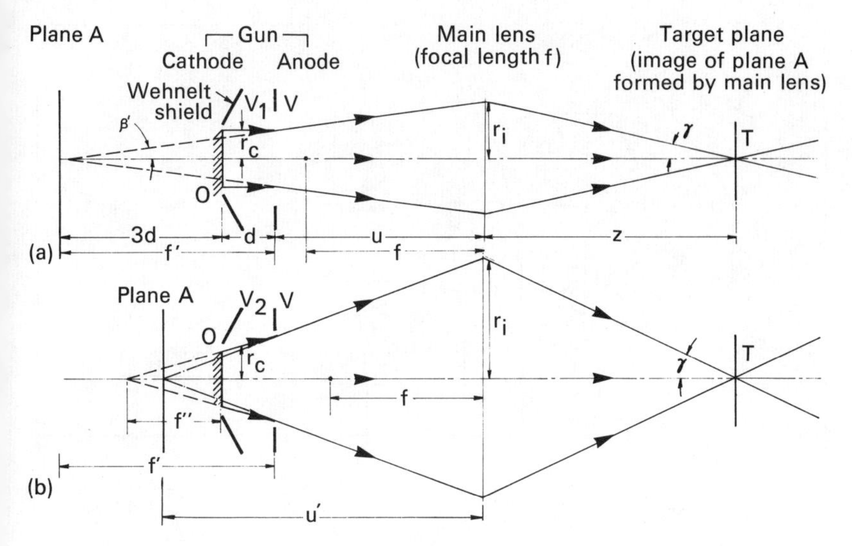

Fig. 2.13 Electron-beam systems: (a) parallel (cylindrical) beam in gun; focal length of anode aperture diverging lens = f', and (b) diverging beam in gun; focal length of anode aperture diverging lens = f'; focal length of equivalent additional diverging lens at cathode = f''

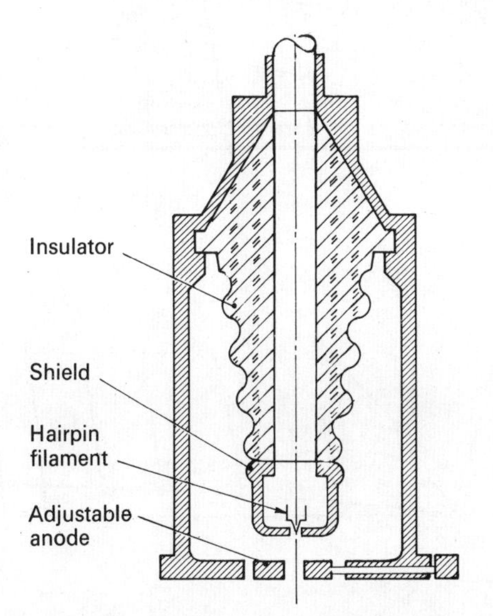

Fig. 2.14a Typical 100kV electron microscope gun (schematic) (*Courtesy GEC/EE/AEI*)

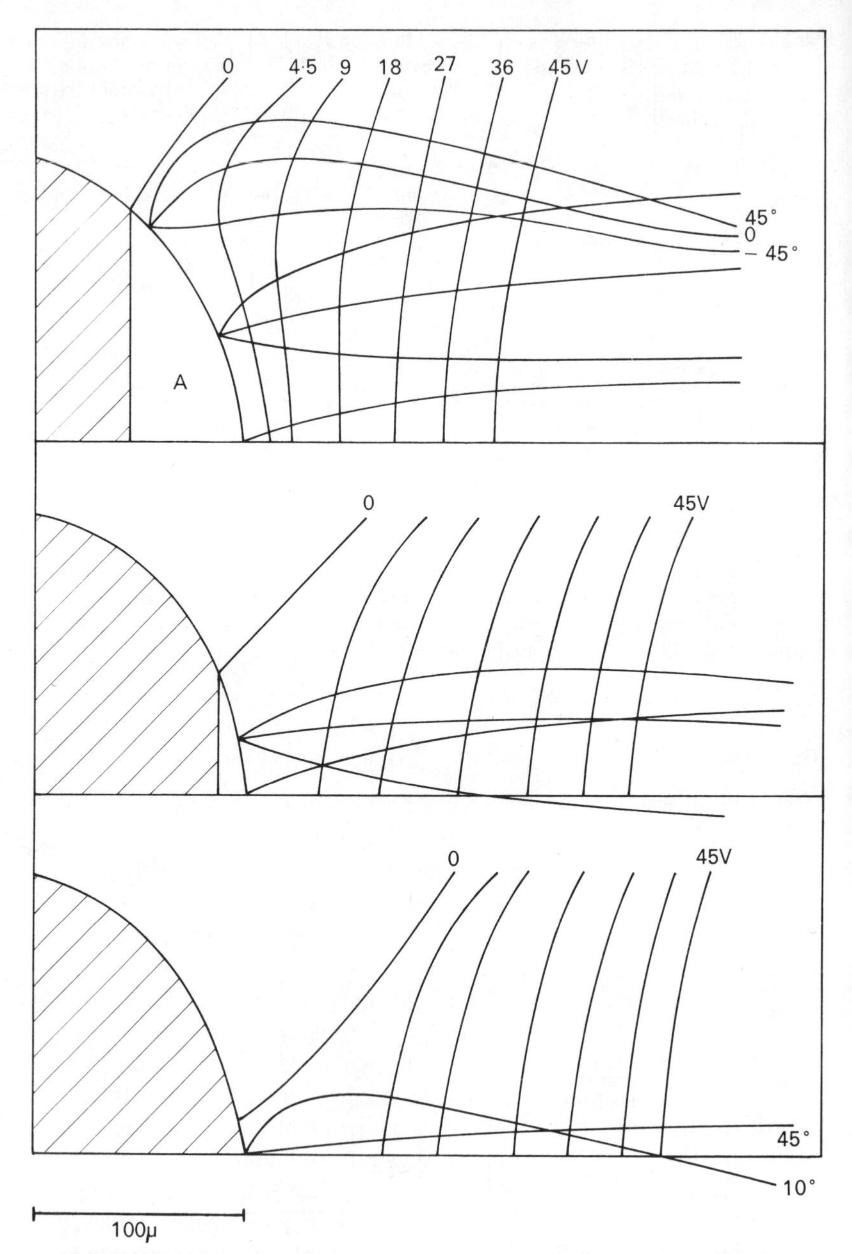

Fig. 2.14b Equipotentials and electron trajectories near hairpin filament tip as bias is increased towards 'cut-off'.

Anode potential = 50kV. A = emitting region

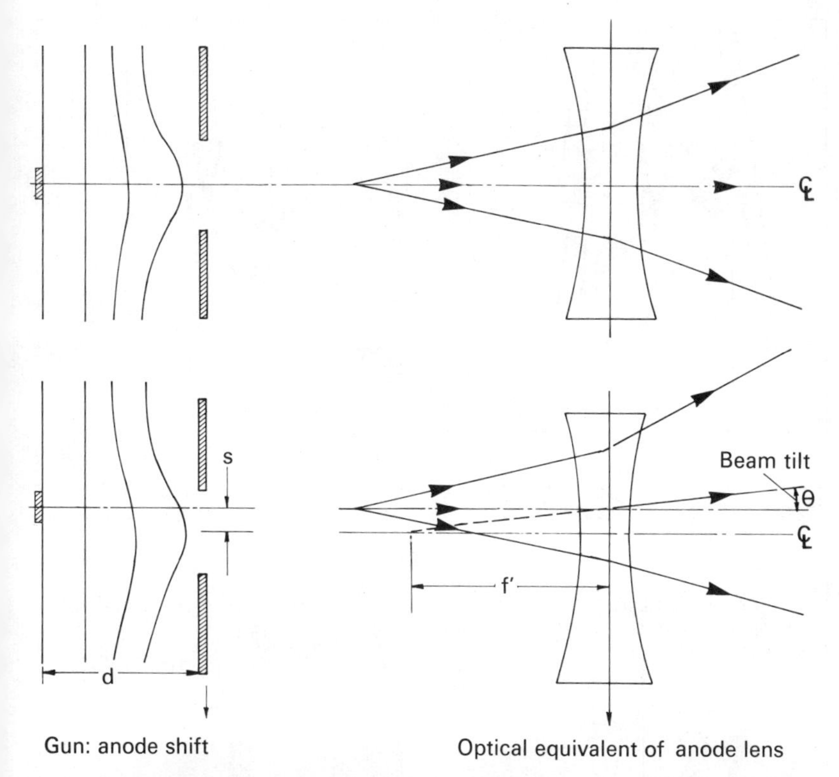

Fig. 2.15 Lateral displacement of anode plate with its optical equivalent

can be tilted and thus centred over the main lens by laterally displacing the anode plate (Fig. 2.15). The amount of tilt produced (θ) (radians) is approximately $s/f' = s/4d$ where s is the shift and d the anode/cathode spacing.

Figure 2.16 shows the Steigerwald and the Rogowski types of gun structures. Steigerwald employs the telefocus cathode shield electrode. The Hamilton Standard machines also use hairpin filament cathodes, and in one of their designs the whole gun is tilted through an angle (some 30°) and the beam deflected magnetically into the electron-optical column. This prevents ions, emanating at the target, from hitting the cathode. This design is shown in Fig. 2.17a. The electrode structure of the Hawker Siddeley 150kV, 6kW gun is shown in Fig. 2.17b. This also employs a directly heated filament emitter and a telefocus electrode structure. The whole

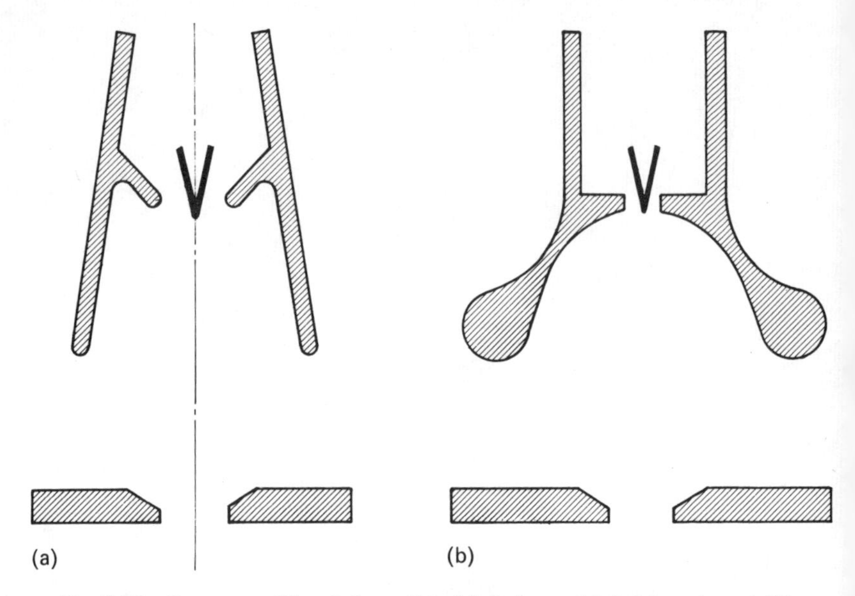

Fig. 2.16 Gun assemblies (schematic): (a) Steigerwald (telefocus), and (b) Rogowski

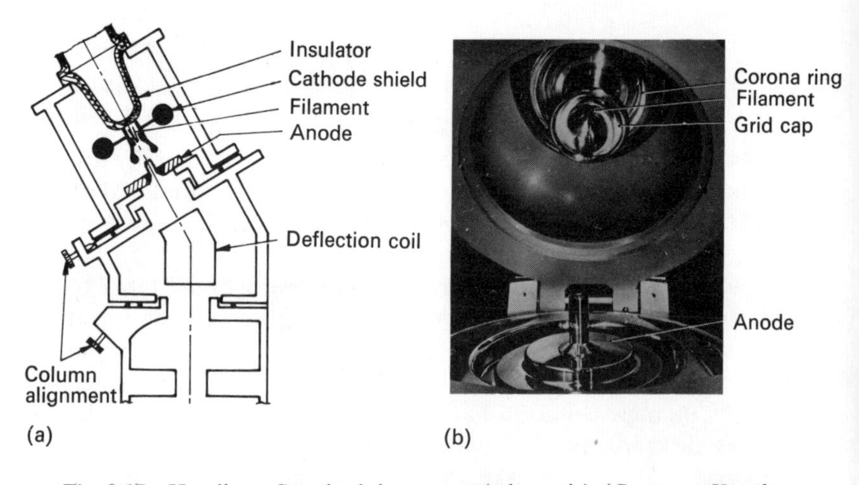

Fig. 2.17 Hamilton Standard bent gun (schematic) (*Courtesy Hamilton Standard Division of United Aircraft Corporation*), and Hawker Siddeley 150kV, 6kW gun (*Courtesy Hawker Siddeley Dynamics, Electron Beam Division*)

cathode and high-voltage assembly is hinged to give ready access to the electrodes for replacing cathodes.

A diode type of gun operating on the Pierce cylindrical-beam system is the Sciaky design shown schematically in Fig. 2.18, which operates quite successfully in the range 30–60kV, yielding powers of over 10kW. Its chief limitation is the lack of facility of control or regulation of current other than by variation of anode/cathode voltage or spacing. This design has since been superseded by an improved version provided with bias control of the current and operating at up to 30kW.

INDIRECTLY HEATED CATHODES. The advantages of indirectly heated cathodes are longer life, greater source stability, and better defined spots. Considerable work has been done by Bas[22] who has developed a gun in which the cathode is a cylindrical tungsten rod, emitting

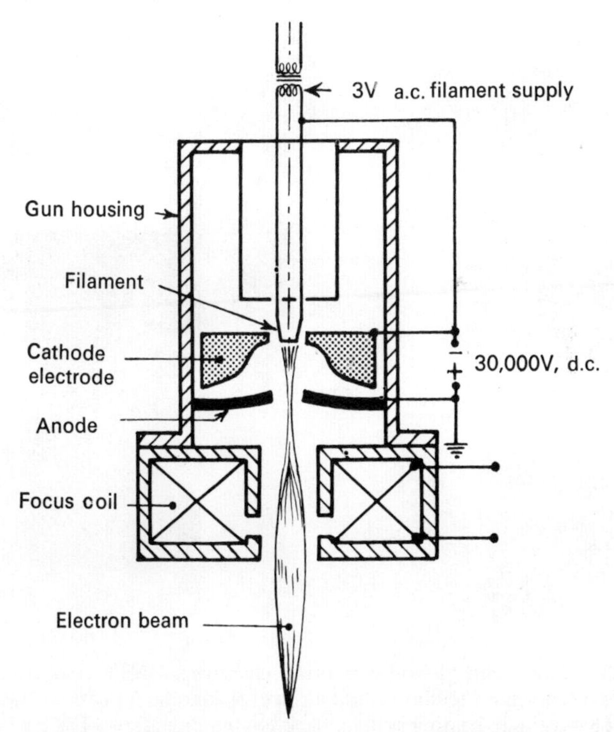

Fig. 2.18 Sciaky gun (*Courtesy Sciaky Electric Welding Machines Ltd*)

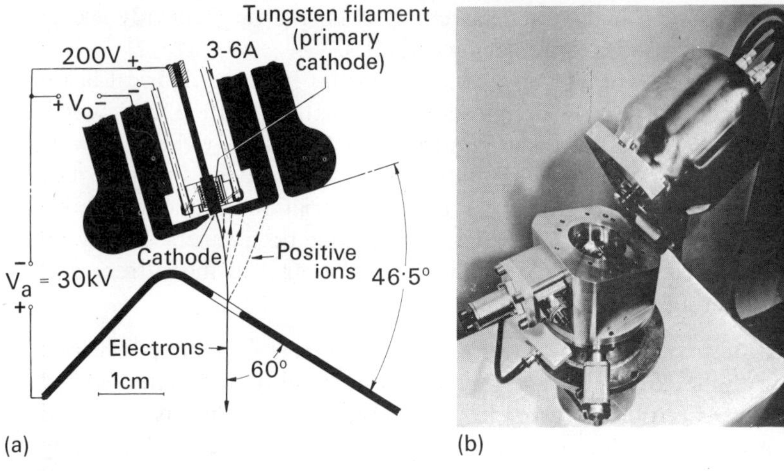

Fig. 2.19 Bas gun (*Courtesy E. B. Bas*)

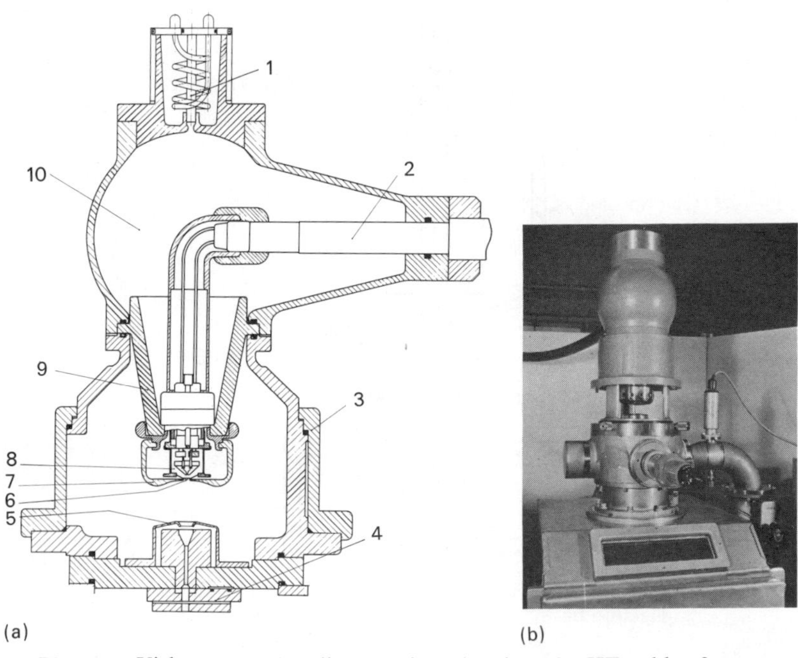

Fig. 2.20 Vickers gun: 1—oil expansion chamber; 2—HT cable; 3—gun access cover; 4—gun vacuum-isolation valve; 5—anode; 6—cathode button; 7—grid; 8—cathode-heating system (back bombardment); 9—HT insulator; 10—oil insulation (*Courtesy Vickers Ltd*)

[Approx. scale: Access cover (3) dia. = 30 cm.]

from its end face, and which is heated by electron bombardment (see Fig. 2.19). The gun works on the triode principle and has been incorporated in a machine of 5kW total power.

Extensive development of the back-bombarded welding gun has also been carried out by Vickers Ltd. The design is shown in Fig. 2.20 and schematically in Fig. 2.21. The shape of the shield approximates to that required to give a space-charge limited conical beam at zero bias, and is determined by the procedure outlined by Spangenberg *et al.*[9,12] Biasing then provides current control. The back-bombarding or auxiliary electron gun is carefully designed to attain maximum economy in heating the button. The cathode button is made from tantalum and calculations accounting for the balance of heat flow from the button by conduction and radiation show a requirement of button input power of about 80W to raise it to the

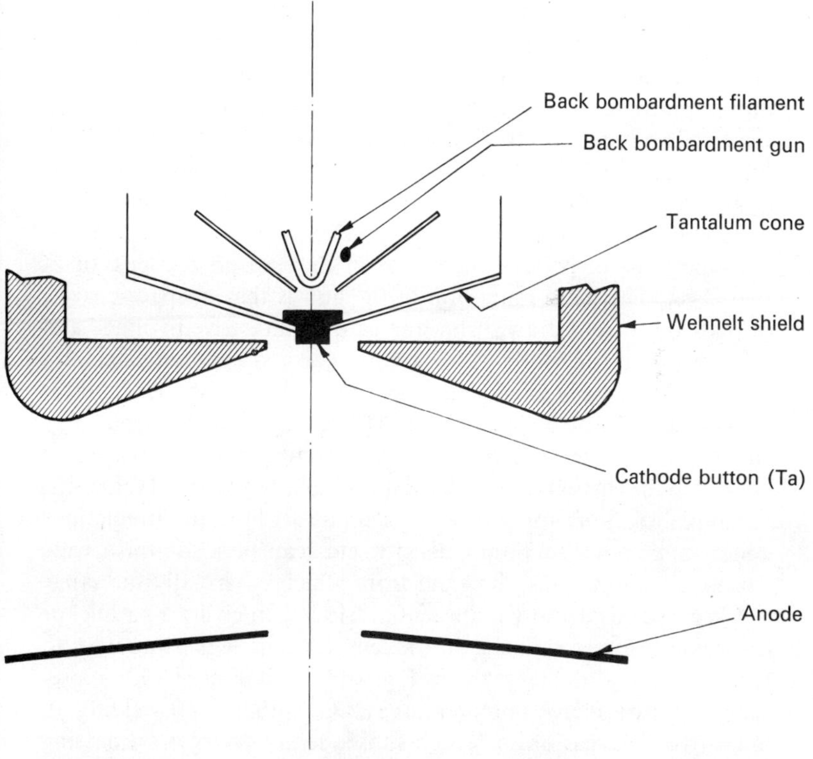

Fig. 2.21 Back-bombardment heating system (schematic)

required temperature of 2500°K for a button of 1·5mm diameter. This power is provided by a current-stabilized back-bombarding system operating at 4kV and 20mA. The cathode is supported on a conical holder and this assembly almost completely surrounds the back-bombarding auxiliary gun (Fig. 2.21). No electrons from this auxiliary system can escape into the main beam, where they could give rise to unwanted heating effects as well as deterioration of spot size.

The function of the oil in the cable chamber is primarily to provide high-voltage insulation but it also serves to extract the heat generated by the bombarding system from the gun electrodes.

Beam focus

Focal limitations

Since the function of an electron system designed for EBW is the attainment of a high power density electron spot at the target, the entire beam-forming system must be considered rather than the gun in isolation. Although a number of lenses may be used to form the electron image at the workpiece, it soon becomes clear that one imaging lens is adequate; no fundamental improvements will accrue apart from that of practical convenience, as, for example, the production of intermediate images in differentially pumped systems.

Limitations of performance arise in the gun, in the lens, or at the target. The main limitation in the gun is that of space-charge, in that the current drawn from the cathode at a given voltage and for given electrode configurations cannot exceed a certain maximum value. The lens may introduce aberrations, chiefly spherical aberration, which will degrade the spot. At the target, the electrons may suffer space-charge spreading at high current densities and are also subject to the current-density limitation caused by thermal velocities.

Simple relations apply to the system as a whole and predictions concerning spot size, power density, etc., can be made for a wide variety of gun designs, the yield from which is virtually the same.

We now consider any unipotential triode gun giving a parallel or diverging beam of radius r_i in the lens, a single lens being used to form the spot at the target as in Fig. 2.13a or b. Figure 2.13a represents the pure Pierce unipotential system with $V_1 = 0$ yielding an initially cylindrical beam, which subsequently diverges on passing through the anode aperture whose lens action has a focal length

$f' \simeq 4d$; Fig. 2.13b shows a more general case of a Pierce-type triode gun where the potential on the shield is V_2 (other than zero) thus introducing a further focusing action in the gun equivalent to the inclusion of an additional diverging (or converging) lens at the cathode of focal length f''. In both cases, the main lens forms an image of plane A at the target T.

Effect of thermal velocities

Electrons are emitted from a cathode with a distribution ranging from zero to infinity in energy, and over a solid angle of 2π given by the well-known Maxwellian distribution formula:

$$\mathrm{d}N(E, \theta) = N\frac{E}{kT}\exp\left(\frac{-E}{kT}\right)\frac{\mathrm{d}E}{kT}\cos\theta\,\mathrm{d}\theta \qquad (2.10)$$

where $\mathrm{d}N(E, \theta)$ is the number of electrons having energies lying between E and $E + \mathrm{d}E$, and which are emitted between the angles θ and $\theta + \mathrm{d}\theta$.

N is the total number of electrons emitted
θ the direction of emission to the normal to the emitting surface

It is the existence of this finite spread in energy which is a fundamental limitation to the attainment of infinite current density in a focused spot. In considering this spread, Langmuir[23] (see also Moss[24]) has derived the relation between current densities (ρ_c and ρ_i) in cathode object and image space, in a system in which a cathode is imaged by a lens (Fig. 2.22). For small magnifications M, the ratio of the two current densities is given by the familiar expression

$$\frac{\rho_i}{\rho_c} \simeq \left(1 + \frac{eV}{kT}\right)\sin^2\alpha \qquad (2.11a)$$

where V is the anode voltage, kT is the most probable emission energy ($= V_0 e$) and α is the arrival beam semi-angle at the image. The maximum current density possible in any cathode image is therefore:

$$\rho_{i\max} = \rho_c\left(1 + \frac{eV}{kT}\right) \qquad (2.11b)$$

at zero magnification and large beam angle α.

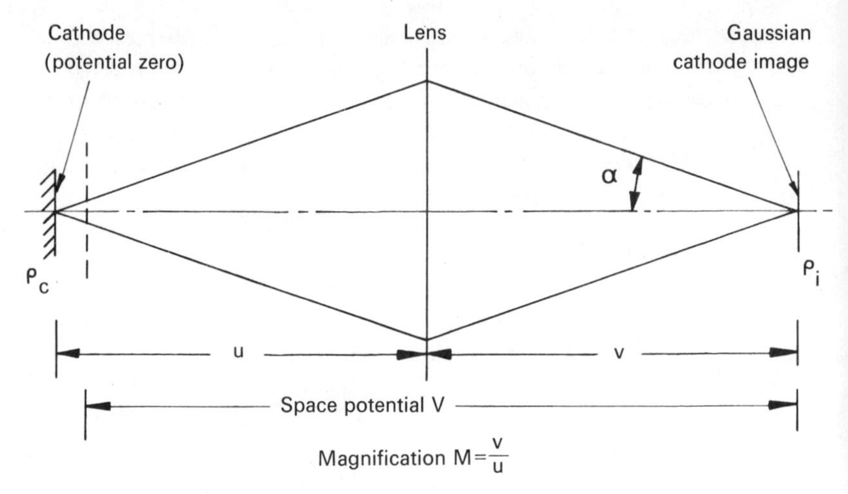

Fig. 2.22 Cathode imaged by lens

In a system such as that shown in Fig. 2.23, even in the absence of any space-charge effects, the electrons would not be concentrated into a point but into a disc of minimum size whose radius is given by

$$r_{\mathrm{T}}\left(=\frac{\delta_{\mathrm{T}}}{2}\right) \simeq \frac{2r_{\mathrm{c}}}{(\sin 2\theta)\sqrt{\dfrac{V}{V_0}}} \tag{2.12a}$$

$$= R_{\mathrm{C}}\sqrt{\frac{V_0}{V}} \tag{2.12b}$$

It is seen that eq. (2.12b) can be derived directly from eq. (2.11a) since

$$\rho_{\mathrm{i}} = \rho_{\mathrm{c}}\left(1 + \frac{V}{V_0}\right)\sin^2\theta$$

or

$$r_{\mathrm{T}} = r_{\mathrm{c}}\,(\sin\theta)\sqrt{\frac{V}{V_0}}$$

In considering the spreading of the spot due to the thermal velocities alone, it is necessary to examine the general ray diagram applying to both unipotential and triode guns (Fig. 2.13) in some more detail. The target (plane Z) is placed at the image of plane A formed by the main lens.

A spot of minimum diameter approximately δ_T ($= 2r_T$) only can be attained and this, by applying the Langmuir relation [eq. (2.11a)] can be derived simply as follows:

$$\rho_T = \rho_c\left(\frac{V}{kT} + 1\right)\sin^2\gamma \text{ (}V\text{ and }kT\text{ in volts)}$$

or

$$\frac{r_T}{r_c} = \sqrt{\frac{\rho_c}{\rho_T}} = \sin\gamma\sqrt{\frac{kT}{V}} = \frac{z}{r_i}\sqrt{\frac{kT}{V}} \tag{2.12c}$$

(z = 'throw distance' or 'working distance' = distance from lens to target.)

Hence $$\delta_T = 2r_T = \frac{2r_c}{r_i}z\sqrt{\frac{kT}{V}}. \tag{2.12d}$$

If only a portion r_e of the cathode is emitting then r_c should be replaced by r_e in eq. (2.12d).

It is significant to note that this thermal limitation depends, for a given throw distance, directly on the cathode radius; therefore, to make δ_T small, the cathode radius should be reduced. Naturally a compromise will have to be struck between the minimum r_c and the requirement for emission current.

Space-charge spreading at the target

Electrons converging in a conical beam to a point at the target will be subject to mutual repulsion which limits the minimum spot size attainable. The repulsion becomes increasingly severe with increasing current in the beam, so that the condition varies from that of the purely conical beam at low current to that shown in Fig. 2.24 for high current content, where the electrons are actually prevented from crossing the axis. For intermediate currents, some hybrid conditions arise which are not readily calculable. At present, however, our interest centres around the space-charge limitation imposed at high currents. A universal space-charge spreading nomograph giving the desired minimum spot size $2r_s$ at the target has been derived by Hollway.[25] Alternatively, an approximate formula for the minimum spot radius at the target may also be applied:

$$r_s = 5{\cdot}9 \times 10^4 \times Z^{5/2} \times I^{5/4} \times V^{-15/8} \times r_i^{-3/2} \tag{2.13}$$

where distances are in cm, potential in volts, and current in amperes. At voltages above about 100kV, the relativistic correction $V_r = (V + 9{\cdot}74 \times 10^{-7}V^2)$ is significant and should be included.

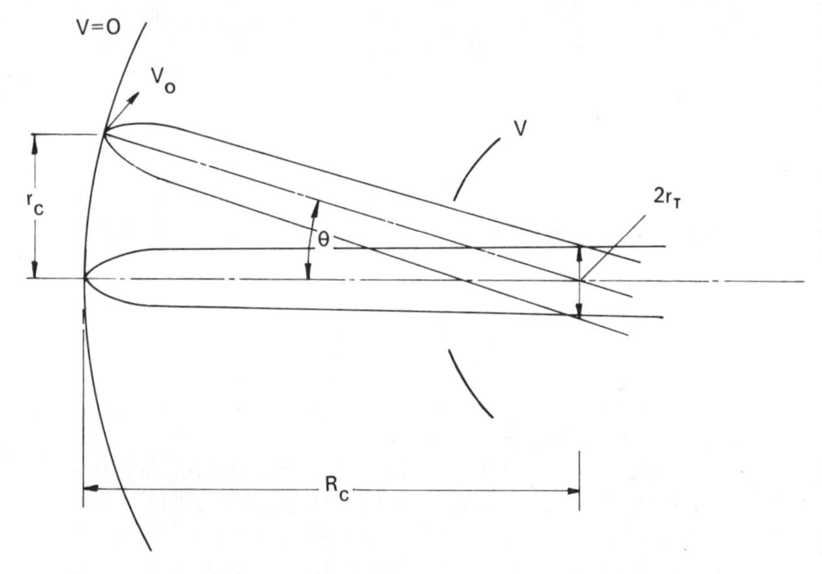

Fig. 2.23 Electrons concentrate in disc, not spot, owing to thermal velocity effects

Spherical aberration of the focusing lens

Spherical aberration results in the marginal rays crossing the axis at a different position from the paraxial rays. This causes an ideal point image to be confused into a disc (Fig. 2.25) whose minimum diameter is given by

$$\delta_{SA} = 2r_{SA} = \tfrac{1}{2}C_S\gamma^3\frac{z^4}{f^4} \tag{2.14}$$

where

C_S = spherical aberration constant of the lens
γ = convergence angle of the electrons
f = focal length of main lens = $(u'z)/(u' + z)$ (Fig. 2.13)
u' = 'object distance' from plane A to lens.
z = 'throw distance' = distance from lens to target.

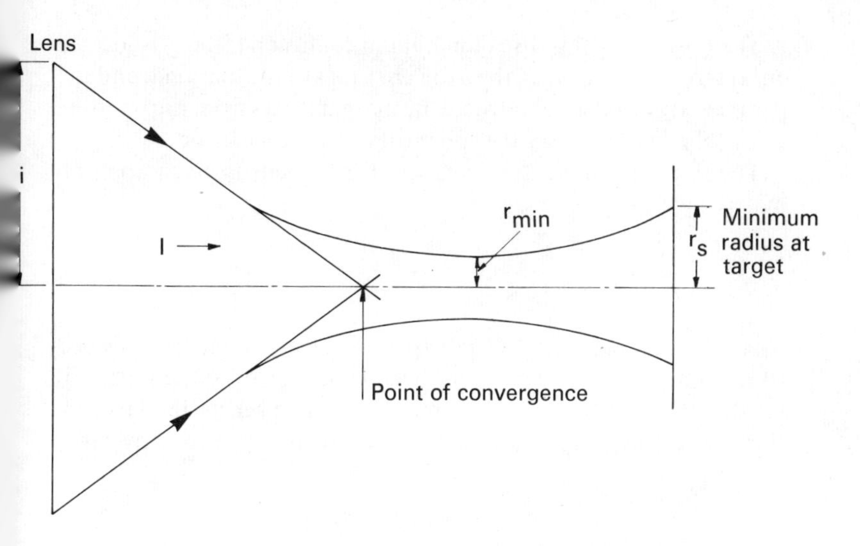

Fig. 2.24 At heavy currents, space-charge spreading prevents electrons from crossing axis

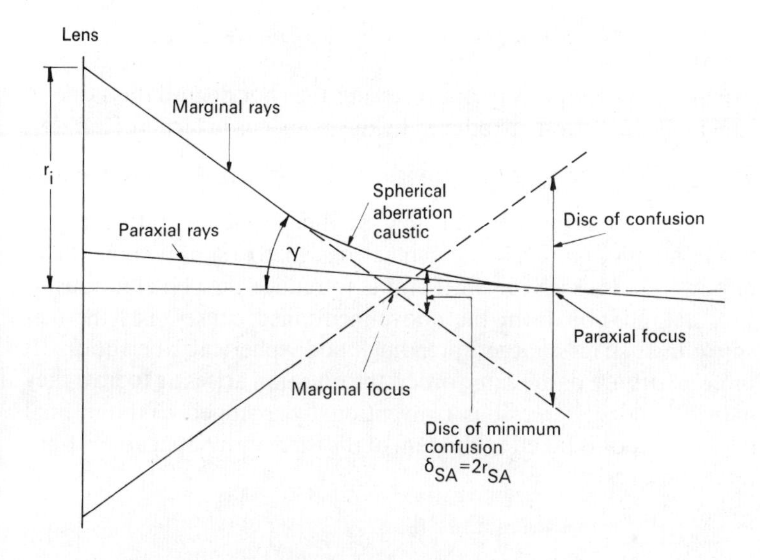

Fig. 2.25 Spherical aberration

The position of this disc of minimum confusion ($\delta_{SA\,min}$) is located one quarter way between the axial crossings of the marginal and the paraxial rays, at the point where the marginal rays from the opposite side of the lens intersect the spherical aberration caustic.

The spherical aberration constant for the lens is given approximately by[26,27]

$$C_S = \frac{5f^3}{(S + D)^2} \tag{2.15}$$

where S and D are the lens pole-piece separation and bore respectively in the case of the magnetic lens. Thus, C_S can be minimized by making $(S + D)$ large, but a limitation is here set by the fact that there is a minimum value of focal length possible for any given $(S + D)$. This is given by:

$$f_{min} \simeq 0{\cdot}5(S + D) \tag{2.16}$$

This means that, if short throw distances z and thus short focal lengths are required, then $S + D$ also must be made small.

By combining eq. (2.14), (2.15), and (2.16), we arrive at

$$\delta_{SA} = 2{\cdot}5r_i^3 \frac{z}{f(S + D)^2} = 2{\cdot}5r_i^3 \left(\frac{u' + z}{u'}\right) \frac{1}{(S + D)^2} \tag{2.17*}$$

which gives a useful way of expressing the spot degradation due to spherical aberration introduced by the main lens in terms of z and r_i.

Practical beam design

The practical approach will now be illustrated for a Pierce-type triode gun plus one lens arrangement, designed to supply some 10kW of power at 150kV. The chief aim is to reduce the contributions to spot degradation from the above-mentioned causes, i.e., thermal spreading, space-charge spreading, and spherical aberration. If these errors are of the same order, the question arises as to how they may be added. Here it is conventionally assumed that the final degraded spot is taken as the sum of the three individual aberrations added in quadrature:

$$\delta = \sqrt{(\delta_T^2 + \delta_S^2 + \delta_{SA}^2)} \tag{2.18}$$

* Strictly true only for a weak lens.

It should be realized that, because of the complicated current density distribution at the spot, such an addition may not be entirely justified and a more detailed analysis of conditions at the spot is necessary. However, this procedure is considered reasonable in the first instance.

The lens design should be chosen initially so that $S + D$ is as large as possible but consistent with the requirement for the focal length $f = 0{\cdot}5(S + D)$, i.e., the maximum value of $(S + D)$ is about equal to the shortest throw distance z and in the present example is made 10cm.

Increasing r_i, through deliberate control of the shape of the cathode shield, will decrease the space-charge contribution at the spot but increase that of spherical aberration. These two effects should be equated at the greatest required throw distance z to fix an optimum r_i. This now allows a determination of r_c from eq. (2.12a), so that r_T does not vastly exceed r_S and r_{SA} and at the same time is consistent with a practical value of cathode emission current density, say, $5A/cm^2$. Attainable target spot diameters and power densities can now be calculated and in the present example the conditions were selected:

$$r_c = 0{\cdot}075\text{cm}$$

$$V = 150\text{kV}$$

$$I = 70\text{mA maximum}$$

with z and r_i as independent variables, plotted in Fig. 2.26. It is seen that, for short throw distances z, spot sizes are much smaller, chiefly because the space-charge effect is small. By increasing r_i, the space-charge spread can be reduced for the longer throw distances z in the range 20–50cm at the expense of the spot size at shorter throw distance. The object distance u need not be increased physically for this to be achieved; the effect can be secured as stated by increasing the initial beam divergence through adjustment of bias or by shaping the cathode profile. Similarly, an auxiliary lens close to the anode would achieve the same effect and allow control of r_i.

The predictions embodied in the curves of Fig. 2.26 are substantiated reasonably well in practice. The results of some tests on the Vickers electron gun illustrated in Figs. 2.20 and 2.21 are shown in Fig. 2.27. Practical measurements of spot diameters have been made by the technique described later and are compared with the present calculations to obtain Fig. 2.27. The theoretical curves here have been derived from the curves of Fig. 2.26 and plotted with beam current as abscissa, spot size as ordinate, and throw distance as a fixed parameter. These particular measurements were made with an unstablized power supply; therefore allowance has been made for the voltage droop of the power supply at high current, which makes the estimate of the space-charge effect at long throw distance somewhat more realistic.

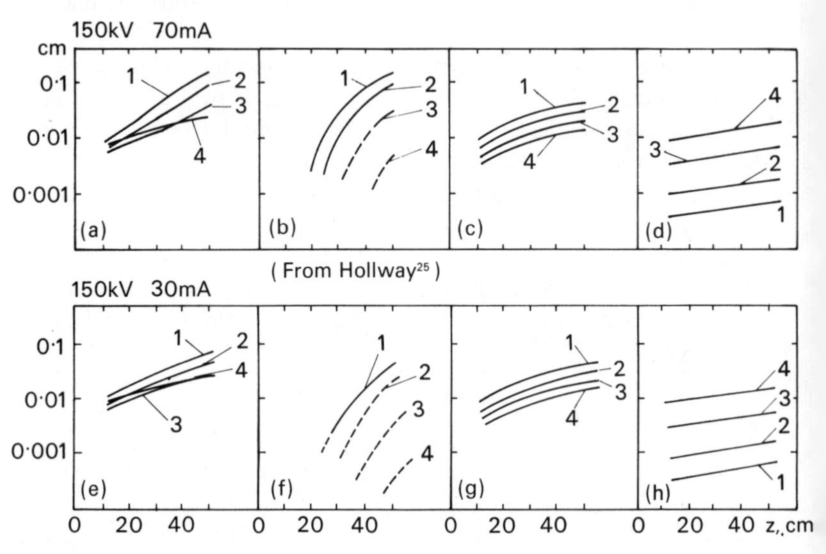

Fig. 2.26 Theoretical spot sizes. Key: 1: $r_i = 0{\cdot}225$cm; 2: $r_i = 0{\cdot}300$cm; 3: $r_i = 0{\cdot}450$cm; 4: $r_i = 0{\cdot}625$cm. (a) $\delta = \sqrt{(\delta_S^2 + \delta_T^2 + \delta_{SA}^2)}$, (b) $\delta_S = 2r_S$, (c) $\delta_T = 2r_T = 2z\alpha\frac{r_c}{r_i}$, (d) $\delta_{SA} = 2r_{SA} = 2{\cdot}5r_i^3\frac{u' + z}{u'}\left(\frac{1}{S + D}\right)^2$, (e) δ, (f) δ_S, (g) δ_T, and (h) δ_{SA}

The design for operation at other voltages and powers is exactly similar. Some calculations[28] have shown that spot power density can certainly be markedly increased at high currents, particularly when the voltage is also raised, or when working at shorter throw distances (below 20cm). On the other hand, for the long throw distances, around 50cm, the space-charge spreading negates any gain resulting from an increase in power, and power density falls off very rapidly.

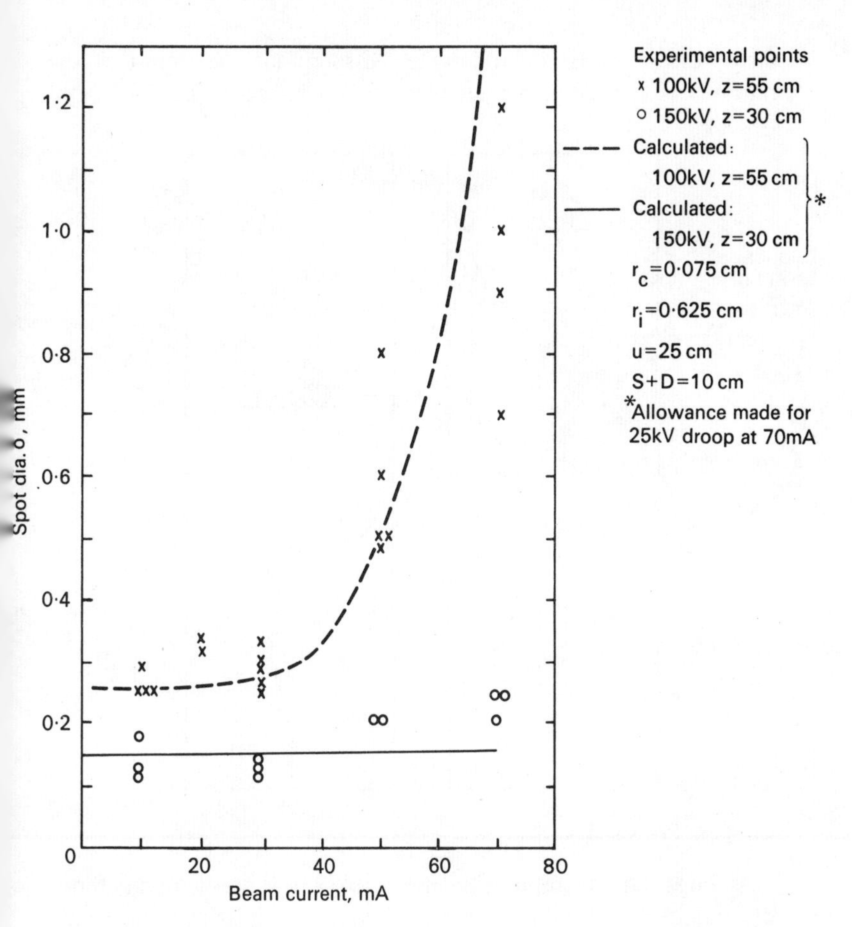

Fig. 2.27 Calculated spot diameters and experimental points obtained with Vickers electron gun

Two-lens system

In special instances it is sometimes convenient to employ more than one focusing lens. In general, there is no electron-optical advantage in so doing as it is always possible to replace a number of lenses by a single equivalent lens, appropriately positioned along the beam axis. However, where flexibility of the electron optics of the column is required, two lenses are a practical help, e.g., if the column is to be extended in length to allow for optical viewing or for the insertion of a deflection system, or if the high power density associated with short throw distances is to be retained throughout the depth of the chamber (Fig. 2.28a) without vertical movement of the whole gun system.

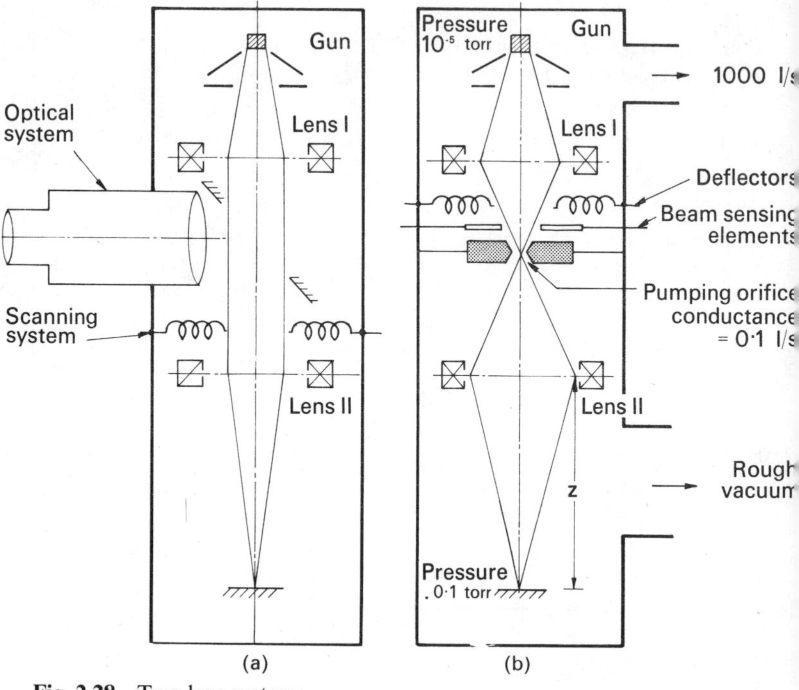

Fig. 2.28 Two-lens systems

An important example of the use of two lenses occurs in differentially pumped EBW systems. In such systems, the aim is to isolate effectively the vacuum system of the gun from that of the chamber as far as possible, while still allowing the electron beam to pass unimpeded. Such an arrangement is of value in restricting the more stringent vacuum demands of the gun to the gun chamber alone and permitting relatively high pressure within the work chamber. This practice is now known as welding under 'soft' vacuum.

A convenient way of constructing a differentially pumped system is shown in Fig. 2.28b, where the work chamber is separated from the gun by a small orifice. An auxiliary lens L_{I} is used to focus the beam onto this orifice and allow it to pass into the work chamber. Lens L_{II} is then used in the normal way to produce a focus on the work piece.

This arrangement is of particular importance, even if the benefits of differential pumping are not the first concern, as it allows a measure of beam position and focus control not possible with a single lens. The first focus formed at the orifice is available for continuous measurement. By the use of beam sensors (Fig. 2.28b) which provide control signals for the lens and deflectors, it is possible to servocontrol the beam position and focus within the orifice to a high order of accuracy. This focus is then independent of beam energy and current, bias potential, cathode wander, etc., and is a positionally stable electron source which is then imaged by L_{II} into the work space. In addition to providing a stable beam position within the work space, this arrangement also confers the benefit that the control of L_{II} may be calibrated and preset in terms of the working distance z for a given beam energy. Such an arrangement would normally be associated with automatic compensation of the lens current for variations of beam energy.

Lens design

The procedure for designing lenses is now fairly well established. For several reasons, magnetic lenses are nearly always used in preference to electrostatic lenses; the latter not only suffer from greater aberrations but also present more difficult practical problems especially at high voltage. The construction of a satisfactory electrostatic lens to operate at, say, 150kV is a considerable engineering feat compared with a magnetic lens.

The need to minimize the spherical aberration introduced by the lens has already been stressed. Spherical aberration has been much discussed in the literature[26,27,29,30] and arises as illustrated in Fig. 2.25; its magnitude is given in eq. (2.15) and (2.17) which also show that, to obtain a minimum focal length, the spherical aberration constant cannot be reduced below a certain value. To design a suitable lens, it only remains to establish the excitation conditions using Liebmann's universally applicable data.[31] His general formula for the focal length of a magnetic lens:

$$\frac{f}{S + D} = \frac{25V}{(NI)^2} \tag{2.19*}$$

* At higher voltages (above 100kV), the relativistic correction $V_r = V + 9{\cdot}74 \times 10^{-7}V^2$ should again be applied.

holds within 5 per cent for S/D ratios varying from 0·2 to 3·0, where

$$V = \text{electron accelerating voltage,}$$
$$NI = \text{ampere-turns of lens winding}$$

Thus, as an example in an EBW machine with a minimum working distance of 10cm ($f \simeq$ 5cm), the condition imposed by eq. (2.16) puts $(S + D) = 5/0{\cdot}5 = 10$cm. This immediately fixes the spherical aberration constant C_S as determined by eq. (2.15).

The required maximum ampere-turns NI from eq. (2.19) become:

$$NI_{max} = \left(\frac{25V(S + D)}{f_{min}}\right)^{1/2} \tag{2.20}$$

which for 150kV works out to 2800.

A well-established value for the safe loading of uncooled lens windings excited continuously is 120 ampere-turns/cm^2 so that the bobbin cross-sectional area required is some 2800/1.20 = 23cm^2. The aspect ratio of this cross-section does not matter critically nor what wire gauge is used, provided the ampere-turns are achieved. What must be ensured is that the pole pieces are not run at magnetic saturation which occurs around 18000G for low-carbon steel.

The magnetic field in the gap is given by:

$$Hp \simeq \frac{4\pi NI}{10S} \tag{2.21}$$

and with S about 5cm, NI = 2800 ampere turns, yields $Hp \simeq$ 700G, i.e., well below saturation.

A satisfactory mechanical construction of the lens which follows from electron microscope practice, in which the winding is divorced from the vacuum, is illustrated in Fig. 2.29. A brass pole-piece insert isolates the winding from the vacuum, without affecting the mechanical alignment of the pole pieces. Precision of roundness and location of the pole pieces is ensured by final machining of the lens bore after assembly of the top and bottom plates which constitute the pole pieces themselves.

Measurement and control of the beam parameters

Beam power

The depth of the weld produced by the electron beam is influenced by several factors. All other conditions remaining constant, the

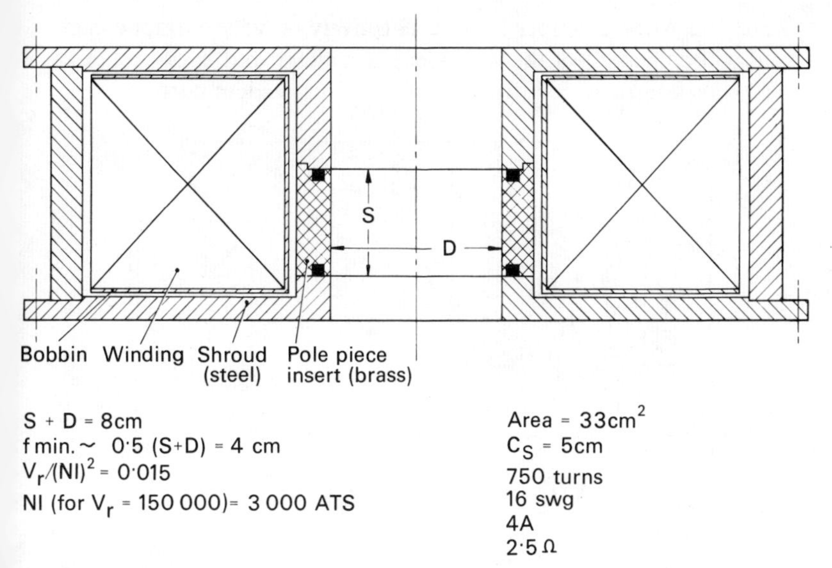

Fig. 2.29 Magnetic lens

depth is directly related to the beam power or, at constant voltage, to the beam current. It follows that this current must be measured and held constant to within a few per cent if consistent welding is to be achieved. This demand is more stringent if full penetration welds in thick sections are to be made. With beam powers of only a few kilowatts, relatively slow welding speeds are necessary, and the difference in beam power leading to incomplete penetration on one hand and 'fall-through' on the other is very small. It is found in practice that the beam current must be held constant to within one per cent. Beam current, in the present context, concerns that portion of the gun emission current which actually reaches the workpiece. A measurement of gun emission current is of limited value as there may be a significant and varying fraction of this intercepted on the anode or other stops within the electron-optical column. Although this could be allowed for in specific instances, it will be found that this allowance varies with changes in cathode position and shape, column alignment, etc. It is naturally more appropriate to measure the current actually arriving at the work, either by insulating the work or the work chamber or, more conveniently, by insulating the normally 'earthy' end of the gun and HT generator as shown in Fig.

2.30. The work chamber is then, effectively, a large Faraday cage and will collect all of the workpiece secondary and back-scattered electrons and provide a true measure of the incident current.

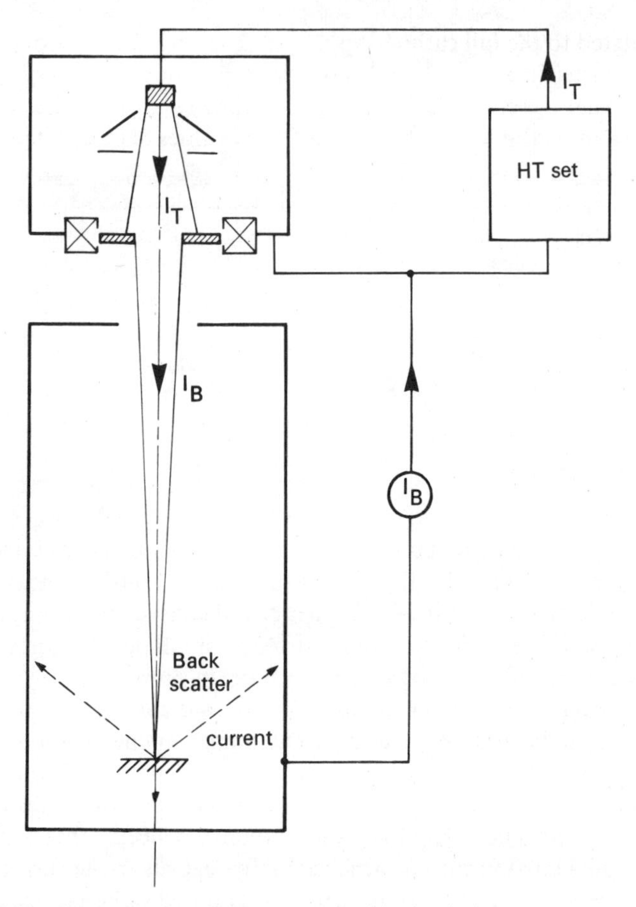

Fig. 2.30 Measurement of current received by workpiece

Control of the current emitted by a three-electrode gun is achieved by biasing, i.e., by making the potential of the cathode shield negative with respect to the cathode. The magnitude of the voltage required for full control depends upon the perveance ($P = I/V^{\frac{3}{2}}$) of the gun and the anode/cathode voltage V. For a typical high voltage EBW gun (150kV, 70mA) with a perveance of 10^{-9}A $V^{-\frac{3}{2}}$, the required control voltage is about $10^{-2} \times V$ (see Fig. 2.12). This increases to around $10^{-1} \times V$ for a perveance of 2×10^{-7}A $V^{-\frac{3}{2}}$, encountered in high-current low-voltage guns.

A simple control system is provided by a dc cathode shield supply, insulated to the full cathode voltage and arranged for the output to be adjusted from the control panel. In practice, such a system is somewhat inconvenient, since slight mechanical drifts of the cathode within the cathode shield call for different values of bias voltage for a given current. Such drifts are always present to some degree, especially during several minutes after the cathode is switched on.

An alternative arrangement is the use of automatic bias.[16] Here the beam current itself generates the bias voltage across a resistor in series with the cathode line. A degree of feedback is thereby provided which helps to stabilize the beam current. However, although simple and effective, it is also not a very convenient system in that adjustments to the current are made by variation of the value of the cathode resistor, which is of course at the full cathode potential.

A more sophisticated system utilizes the work current to servo-control the cathode shield supply (Fig. 2.31). This not only compensates for any drift but also allows the beam current to be accurately preset before the beam is switched on.

A further refinement may be made by feeding this servosystem with a signal proportional to the product of the work current and the cathode voltage, i.e., the work power. If, in this case, the servosystem response time is arranged to be of the order of one millisecond, any transients that might affect the welding process will be suppressed.

When used in conjunction with a lens supply which is compensated for beam voltage variation (focus compensation), there is no longer a need for a highly stabilized power supply to the gun. This is a considerable advantage, particularly at powers above a few kilowatts. These arrangements are shown in Fig. 2.31. It is possible to control this servosystem with signals derived from parameters other than the beam power; e.g., a control signal can be obtained from the electrons penetrating the weld and used to achieve a controlled degree of weld penetration. Such schemes have been contemplated[32] and are now being developed.

To achieve a rapid response, the ac signal, from which the cathode shield voltage is derived, must be high in frequency. A convenient

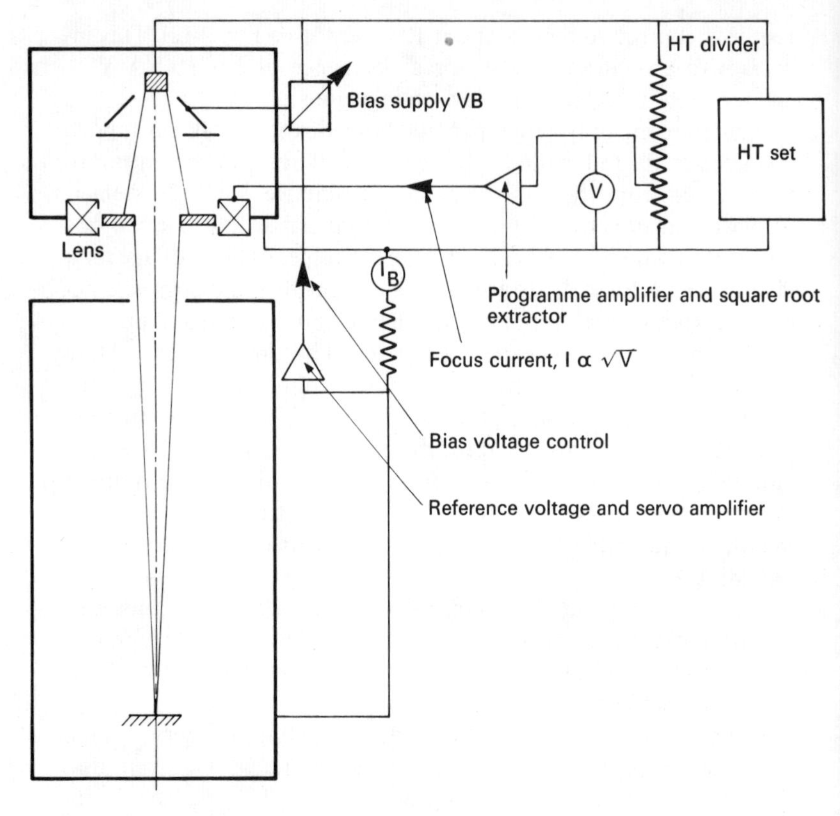

Fig. 2.31 Shield supply controlled by work current

range and one in which suitable high-voltage solid-state rectifiers are available is between 10 and 20kHz. The design of such a power supply and its transformer is very similar to that of any radio frequency driven dc power supply.[33,34]

Power density and distribution

A knowledge of the precise beam profile and power or current density distribution of the beam is required by both the electron-optical designer and EBW plant user. For the unfocused beam, such measurements present no difficulty: the power density is relatively low and it is necessary only that the monitoring system be sufficiently well cooled to deal with the average power involved. A suitable system is shown in Fig. 2.32. The electron beam is scanned across a

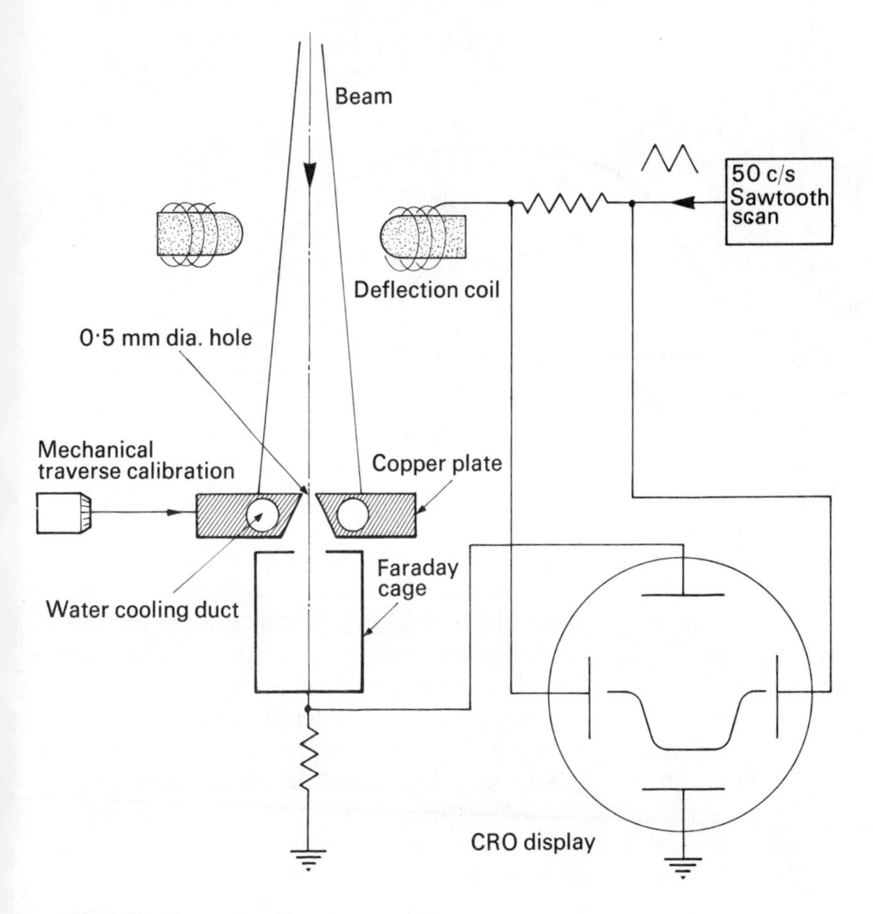

Fig. 2.32 Power density measurement

small hole in a cooled plate; a small fraction of the beam passes through the hole into a current collector and is used to generate a CRO signal so as to display the current density across the beam. A typical distribution is shown in Fig. 2.33.

In the case of the focused beam, the diameter may be extremely small and the power density many orders of magnitude greater than that in the unfocused beam. If the total power is also large—as in many present-day machines—it is scarcely practicable to use the simple system outlined above. The required size of the defining hole or slit is inconveniently small and is subject to large variations

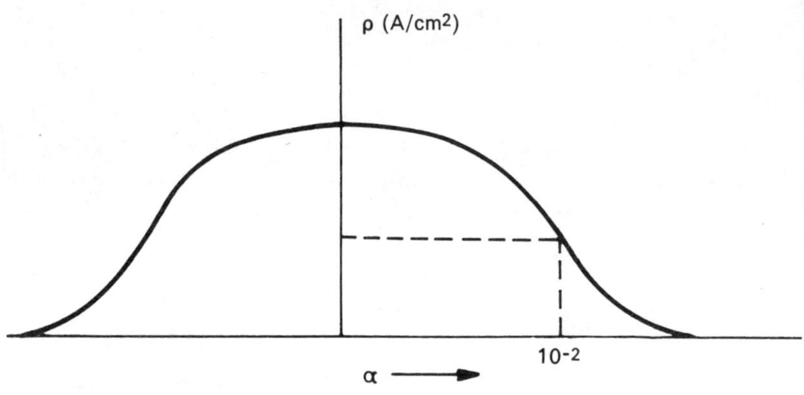

Fig. 2.33 Typical current-density distribution in unfocused lens

because of thermal expansion; there is also the risk of damage by the beam.

A more satisfactory technique for performing such measurements is shown in Fig. 2.34. The focused beam is swept across an edge, and the current which passes the edge at any instant is collected and displayed. The arrangement shown employs two edges separated by a large distance compared to the total diameter of the focused beam. The second edge is of practical convenience in that it provides further protection of the Faraday cage.

The line-scanning rate must be high to minimize local heating and the beam should not retraverse the same path without a suitable time delay, so that adequate conductive cooling may meanwhile take place. A scanning rate of 10cm/s per kilowatt of beam power can be regarded as a minimum for a typical focused beam diameter of 10^{-1}cm.

A further low-frequency scan is applied along the length of the edge so that successive line scans do not overlap. The amplitude of this low-frequency scan should be small to avoid deflection defocusing; alternatively, the Faraday cage should be masked from all but the central region of this scan.

The plates forming the edges are steeply slanted so that the beam is defocused and spread over a large area at the ends of its traverse. This reduces the demand for fast reversal of the scan direction and allows the use of a sinusoidal scan. An alternative mechanical arrangement in which the water-cooled edges may be rotated to expose fresh surfaces is shown in Fig. 2.35.

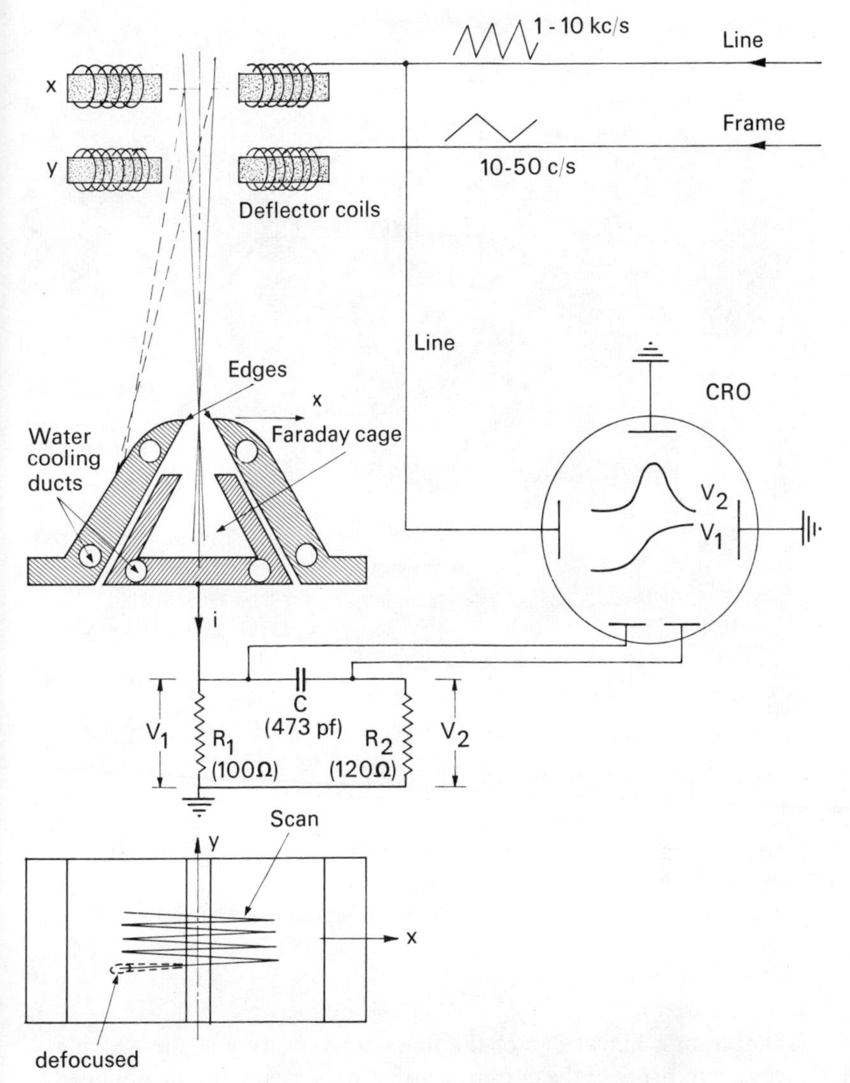

Fig. 2.34 Arrangement for measuring power density distribution

A typical voltage signal $(V_1)_t$ obtained from the Faraday cage is shown in Fig. 2.36a. The total beam current is

$$I_B = i_{max} = \frac{(V_1)_{max}}{R_1}$$

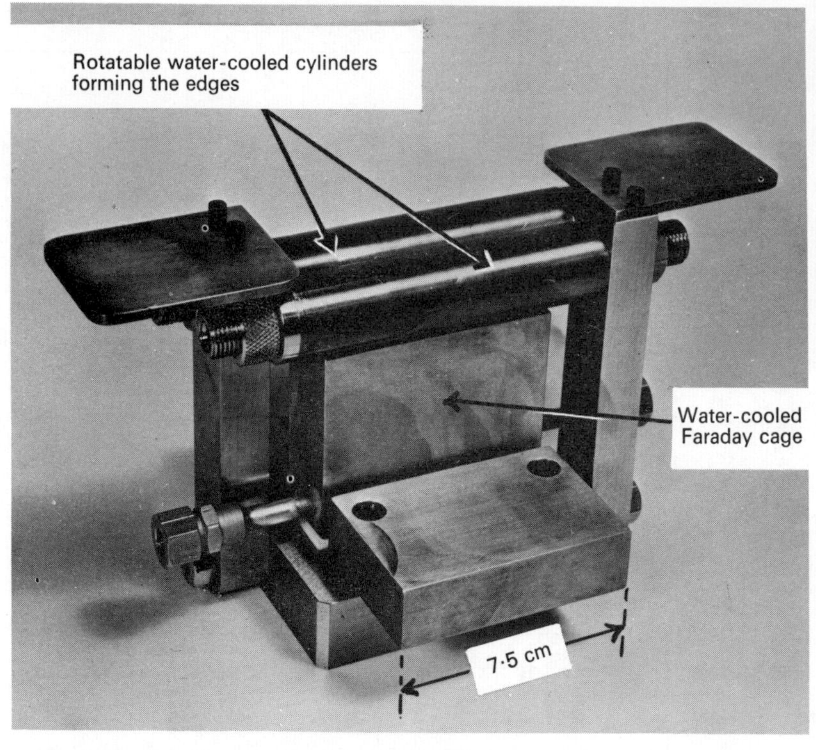

Fig. 2.35 Faraday cage (*Courtesy Vickers Ltd*)

Differentiation of this signal provides the voltage signal $(V_2)_t$ shown in Fig. 2.36b.

If $CR_2 \ll t_1$, the time duration of the signal (see Fig. 2.34), then

$$\frac{(V_2)_t}{CR_1R_2} = \left(\frac{\delta i}{\delta t}\right)_t$$

and, from a knowledge of the line scan velocity v at the defining edge, a measure of the current density in the beam may be obtained

$$\frac{\delta i}{\delta x} = \frac{\delta i}{\delta t} \times \frac{1}{v} = \frac{V_2}{CR_1R_2v} \tag{2.22}$$

This measure is the current per unit width of a strip extending the full width of the beam. A measure of greater relevance is the central current density.

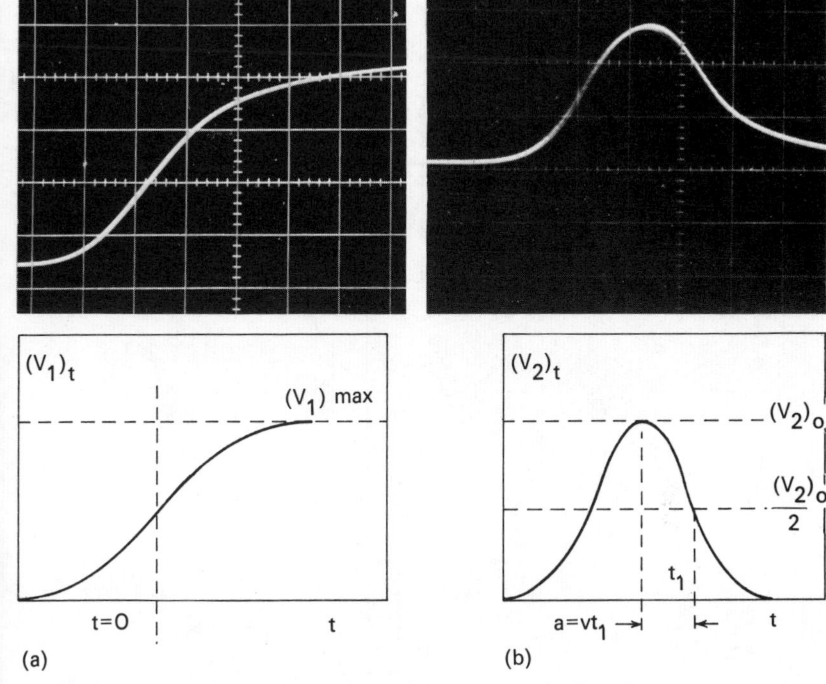

Fig. 2.36 Displayed traces for deriving spot sizes: (a) Faraday cage current, $(i)_t \propto (V_1)_t$, and (b) differential Faraday cage current $(\mathrm{d}i/\mathrm{d}t)_t \propto (V_2)_t$

It may be shown that, if the distribution obtained from eq. (2.22) is gaussian, and if we assume cylindrical symmetry of the beam, then the radial distribution of current density is also gaussian. If the radial distribution of current density is given by:

$$\rho_r = \rho_0 \exp(-kr^2) \tag{2.23}$$

where ρ_0 is the current density at the centre of the spot, and k is a constant, the total current passing the edge at x (Fig. 2.37) is:

$$i_x = \rho_0 \sqrt{\left(\frac{\pi}{k}\right)} \int_x^\infty \exp(-kx^2)\,\mathrm{d}x \tag{2.24}$$

and has the form shown in Fig. 2.36a.

Differentiating eq. (2.24) yields:

$$\left(\frac{\delta i}{\delta x}\right)_x = \sqrt{\left(\frac{\pi}{k}\right)} \rho_0 \exp(-kx^2) \tag{2.25}$$

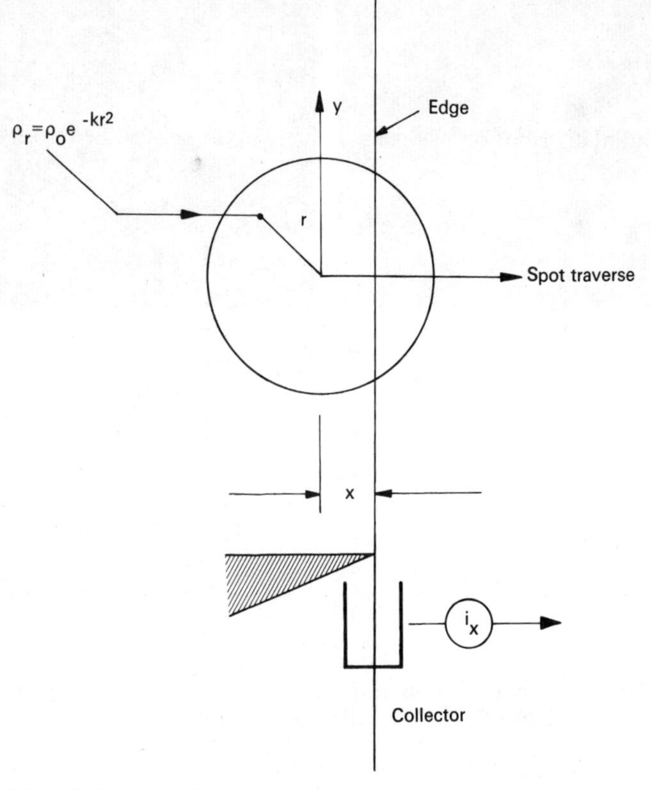

Fig. 2.37 Collector/wedge arrangement

and is the parameter measured from Fig. 2.36b and eq. (2.22), i.e.,

$$\sqrt{\left(\frac{\pi}{k}\right)}\rho_0 \exp(-kx^2) = \frac{(V_2)_x}{v\,.\,CR_1R_2} \tag{2.26}$$

From a measurement of the peak value of $(V_2)_x$ at $x = 0$ and the half-width $x = a$ at $(V_2) = \frac{1}{2}(V_2)_0$ and from eq. (2.26), we may determine the values of the parameters k and ρ_0. Thus, $e^{-ka^2} = \frac{1}{2}$ gives

$$k = \frac{0{\cdot}695}{a^2} \tag{2.27}$$

and by substituting in eq. (2.26) and for $x = 0$,

$$\rho_0 = \frac{0{\cdot}471(V_2)_0}{avCR_1R_2} = A(V_2)_0 \tag{2.28}$$

The radial distribution of current density in the spot is then from eq. (2.23) and (2.28)

$$\rho = A(V_2)_0 \exp(-0{\cdot}695)r^2/a^2 \tag{2.29}$$

It is seen that at a radius r equal to the measured dimension a, the radial current density has fallen to one-half of the maximum value ρ_0. It can also readily be shown that this radius contains 50 per cent of the total beam current. The value of a may therefore be considered as the effective beam radius, and the effective or average power density within this radius is given by:

$$\rho = \frac{VI_B}{2\pi a^2} \tag{2.30}$$

where V is the beam voltage.

Overcurrent and protection

In any demountable high-voltage system, there is always some risk of electrical discharge within the electrode structure. In the event of a flashover between cathode and cathode shield, entailing temporary loss of bias, excessive beam current flows through the column often leading to damage to the workpiece. The frequency with which this occurs depends upon many factors in addition to the basic electrode design. The vacuum conditions, electrode cleanness, and 'conditioning' history, are equally important. Some of the approaches towards eliminating this hazard are mentioned below.

In EBW equipment, air is usually admitted to the work chamber on insertion and removal of the workpiece. It is of considerable advantage to maintain the gun chamber under vacuum during this operation; this may be achieved simply by a suitable valve at the beam output port of the gun column. A separate gun-pumping system with restriction between the gun and the work chamber also provides better vacuum conditions in the gun during welding, thus aiding stability. Other attempts to improve gun stability include at least one gun design in which the cathode is shielded from the workpiece by a bend in the electron-optical column (Fig. 2.17a).

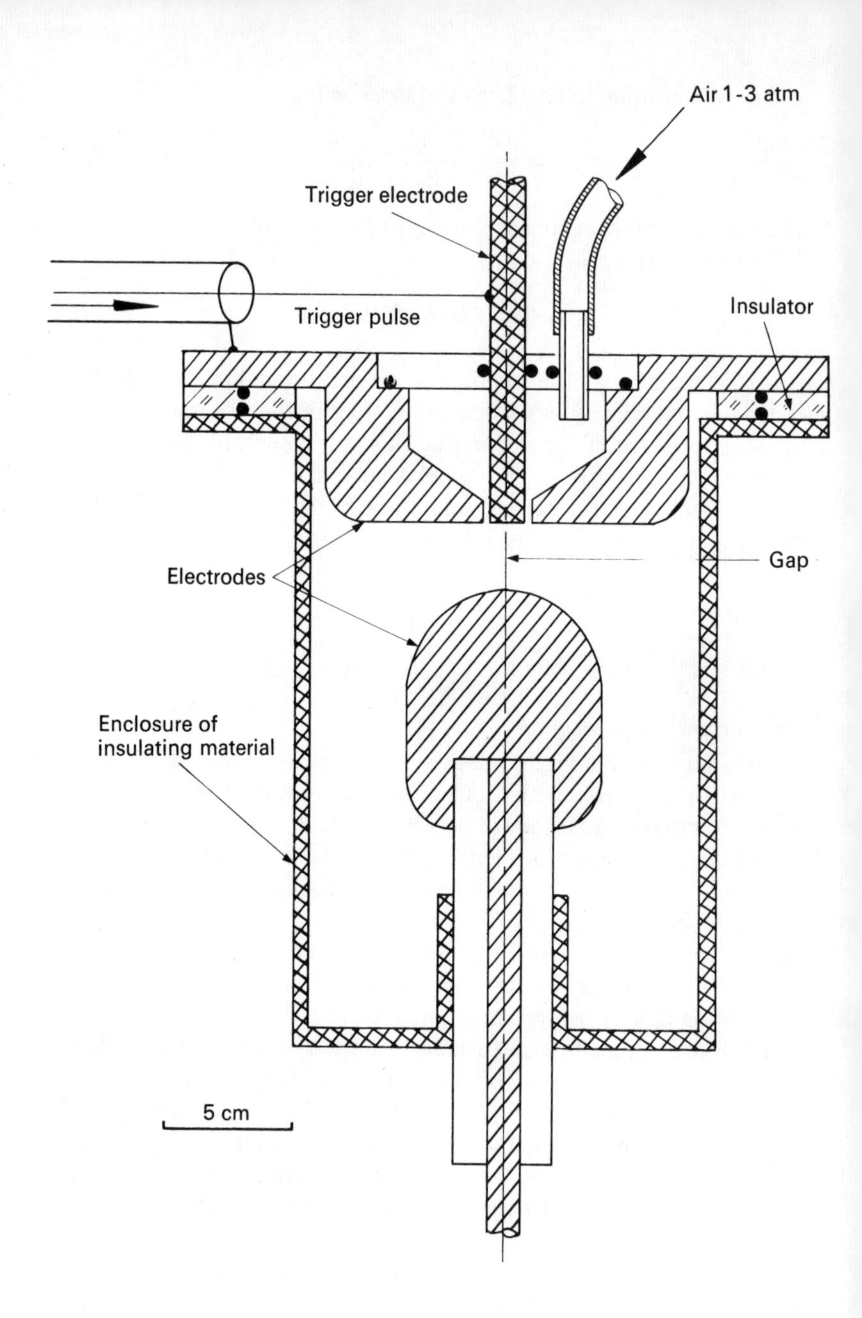

Fig. 2.38 Triggered spark gap (schematic) (*Courtesy Vickers Ltd*)

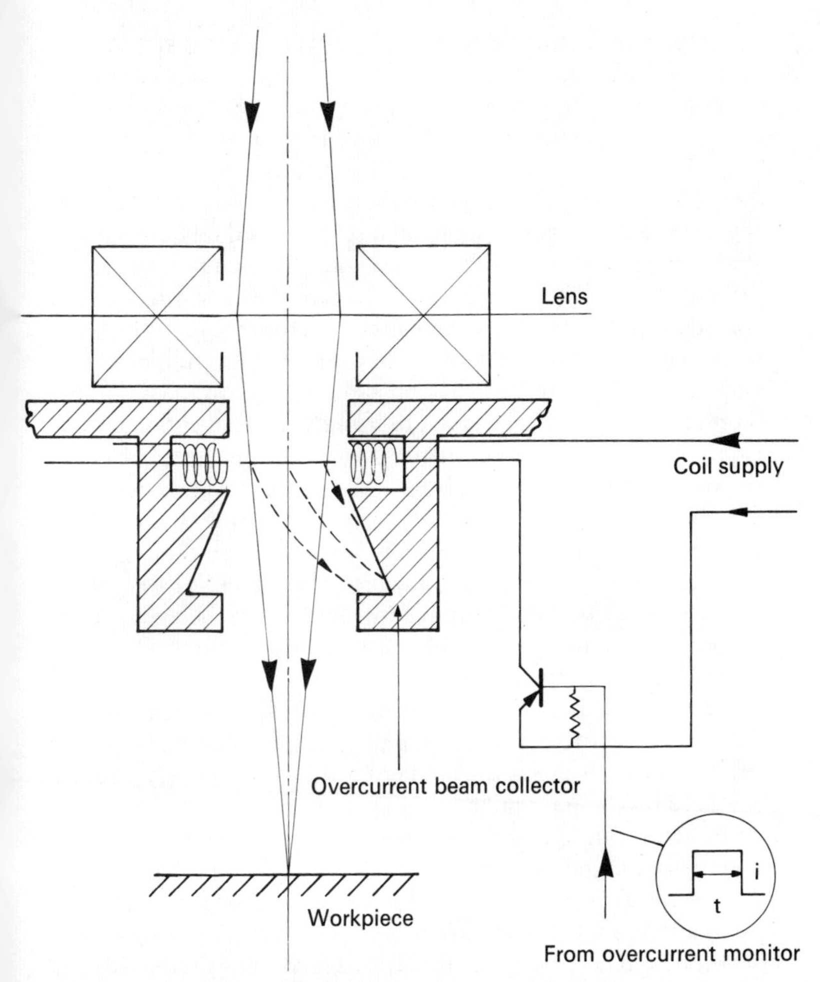

Fig. 2.39 Overcurrent protection

Conditioning the gun to a higher voltage than that to be used during welding is sound practice. Where the high-voltage supply is unstabilized against load, this condition is achieved automatically. Long-life cathodes minimize the necessity to disturb the gun for cathode replacement and avoid the need for frequent reconditioning.

By approaches such as these, the risk of discharge and consequent workpiece damage may be reduced, but not entirely eliminated. The major risk occurs when there is a discharge between the cathode and cathode shield, thereby removing the bias potential and causing maximum beam current to flow. Although the normal overcurrent protection trips will then operate, the interval is relatively long and, in any event, these trips will not remove the stored energy already present in any filter circuits in the HT generator.

There are various ways in which the work may be protected from damage because of these occasional discharges. All rely upon eliminating the beam quickly so as to minimize the energy input to the work during these faulty conditions. Perhaps the most direct way of achieving this is to place a switch directly across the gun which can short-circuit the gun at the onset of overcurrent conditions. This switch must, of course, be fast-acting and tolerant to the full short-circuit current for the time taken to operate the conventional overcurrent trips. This device—sometimes known as an electronic crowbar—may use an ignition, hydrogen thyratron, or spark switch, as the main switch. When the voltage involved is many tens of kilovolts, a triggered spark switch is an obvious and economic choice.

A diagram of such a switch designed for a 150kV EBW gun is shown in Fig. 2.38. This is supplied with a trigger voltage of a few tens of kilovolts in the event of overcurrent and effectively short-circuits the gun within a microsecond of the event.

An alternative method of protection is to deflect the beam so that the current is removed from the work and is spread over a large area collector. If the beam is moved across the work quickly enough, then there will be no significant damage. The problem is similar to that in the design of a spot-size measuring system and similar spot velocities are required. The system is illustrated in Fig. 2.39; it offers the advantage that, on those occasions when the gun flashover is self-curing and the overcurrent is of short duration, the welding operation need not be interrupted. The beam is simply removed from the work for a millisecond or so, which is perfectly acceptable in the vast majority of welding operations.

REFERENCES

1. RICHARDSON, O. W. 'The distribution of molecules in a gas in a field of force.' *Phil. Mag.*, **28** (5), 1914, 633.
2. DUSHMAN, S. 'Electron emission from metals as a function of temperature.' *Phys. Rev.*, **21**, 1923, 623.
3. DUSHMAN, S. 'Scientific foundation of vacuum technique.' 2nd ed. Wiley, New York, 1962.

4. HONIG, R. E. 'Vapour pressure data for the solid and liquid elements.' *RCA Rev.*, **23**, 1962, 567.
5. BLOOMER, R. N. *Proc. IEE.*, 104B, March 1957, 153.
6. HEAVENS, O. S. *Proc. Phys. Soc. B.*, **65** (394), 1952, 791.
7. CHILD, D. C. 'Discharge from hot CaO.' *Phys. Rev.*, **32**, 1911, 492.
8. LANGMUIR, I. 'The effect of space charge and residual gases on thermionic currents in high vacuum.' *Phys. Rev.*, Series 2, **2**, 1913, 450; also with BLODGETT, K. B. 'Currents limited by space charge between coaxial cylinders.' *ibid*, Series 2, **22**, 1923, 347.
9. SPANGENBERG, K. R. *Vacuum Tubes*, McGraw-Hill, New York, 1948, 173.
10. LANGMUIR, I. and COMPTON, K. T. 'Electrical discharges in gases (Part II).' *Rev. Mod. Phys.*, **13**, 1931, 191.
11. PIERCE, J. R. 'Rectilinear electron flow in beams.' *J. App. Phys.*, **11**, 1940, 548.
12. SPANGENBERG, K. R., FIELD, L. M., and HELM, R. 'The production and control of electron beams.' Fed. Telephone and Radio Corp., New York, 1942; also *ibid*, 'Cathode design procedure for electron beam tubes.' *Elect. Commun.*, **24** (1), 1947, 101.
13. PIERCE, J. R. *Theory and Design of Electron Beams*, 2nd ed. Van Nostrand, New York, 1954.
14. THOMPSON, B. J. and HEADRICK, L. B. 'Space charge limitations on the focus of electron beams.' *Proc. IRE.*, **28**, 1940, 318.
15. SCHWARTZ, J. W. 'Space charge limitations on the focus of electron beams.' *RCA Rev.*, **18**, 1957, 3.
16. HAINE, M. E. and EINSTEIN, P. A. 'Characteristics of the hot cathode electron microscope gun.' *Brit. J. App. Phys.*, **3**, February 1952, 40; also *idem* and BORCHERDS, P. H. 'Resistance bias characteristic of the electron microscope gun.' *ibid*, **9**, 1958, 482.
17. BRICKA, M. and BRUCK, H. *Ann. de Radioelec.*, **3**, 1948, 339.
18. DOSSE, J. *Z. Phys.*, 1940, 530.
19. VON BORRIES, B. *Optik*, **3**, 1948, 389.
20. EHRENBERG, W. and SPEAR, W. E. 'Electrostatic focusing system and its application to fine focus X-ray tubes.' *Proc. Phys. Soc. B.*, **64** (373), 1951, 67–75; also SPEAR, W. E. 'Investigation of electron optical properties of electrostatic focusing system.' *ibid*, **64** (375), 1951, 233.
21. HAINE, M. E. British Patent No. 716716.
22. BAS, E. B., *et al*. 'A 30kV electron beam welding machine with indirectly heated tungsten cathode.' Proc. 1st Int'l Conf. on Electron and Ion Beam Science and Technology. 3–7 May 1964, Toronto, Ed. Bakish, R., Wiley, New York, 1965.
23. LANGMUIR, D. B. 'Theoretical limitations of cathode ray tubes.' *Proc. IRE.*, **25**, 1937, 977.
24. MOSS, H. 'On the limit theory of circular electron beams.' Proc. 4th Symp. on Electron Beam Technology, Ed. Bakish, R., Alloyd Electronics Corp., 1962, 9.
25. HOLLWAY, D. L. 'Design chart for calculating electron beam parameters.' *Electronics*, **35** (7), February 1962, 50, 52.

26. HAINE, M. E. *The Electron Microscope*, Spon Ltd, London, 1961.
27. LIEBMANN, G. *Advances in Electronics*, **2**, Academic Press, New York, 1950, 102.
28. EINSTEIN, P. A. and BEADLE, R. 'The design of high power electron beams for welding.' Proc. 2nd Int'l Conf. on Electron and Ion Beam Science and Technology, New York, 1966, Ed. Bakish, R. *Am. Inst. min. met. & pet. Engrs* (*N.Y.*), **1**, 135.
29. MULVEY, T. 'Electron optical design of an X-ray micro-analyser.' *J. Sci. Instruments*, **36**, 1959, 350.
30. EINSTEIN, P. A., HARVEY, D. R., and SIMMONS, P. J. 'The design of an experimental electron beam machine.' *Ibid*, **40**, 1963, 562; also EINSTEIN, P. A. Proc. Nat'l Conf. on Solid Circuits and Microminiaturization, West Ham, 1963, Ed. Dummer, G. W. A., Pergamon Press, Oxford, 1964.
31. LIEBMANN, G. 'A unified representation of mangnetic electron lens properties.' *Proc. Phys. Soc. B.*, **68**, 1955, 737; also HAINE, M. E. Ref. 26, 9.
32. MELEKA, A. H. and ROBERTS, J. K. Institute of Welding, Spring Meeting, London, 1966.
33. BEADLE, R. and BURDEN, D. F. Vickers Research Establishment, Ascot. Unpublished.
34. MAUTNER, R. S. and SCHADE, O. H. 'Television high voltage RF supplies.' *RCA Rev.*, **8**, 1947, 43.

3. Thermal effects

A beam of electrons is pure electric power; pure, in the sense that it is not carried or conducted by matter, such as a conducting wire, a liquid electrolyte, or an ionized gas. Thus, the kinetic energy in an electron beam is transferred directly to the intercepting material. Once the electrons are brought to rest, the whole of their energy is converted to other forms of energy, depending on the nature of the intercepting material and the characteristics of the electron beam itself. Limiting our considerations to interception by solid matter, the electron beam may generate X-rays or light rays as applied in the cases of radiography and the cathode-ray tube respectively. Diffraction of the beam by solid matter has been utilized in electron microscopy. The amplifier tube makes use of the electrical effects of an electron beam. Particle counters and microprobe analysers are more recent applications of the electron beam.

It is clear that many of our technological advances rely on electron beams, primarily perhaps because such beams are generally easy to generate and produce such a diversity of interesting effects. However, in all the applications of electron beams, some heating takes place in the target and this had always been an undesirable feature requiring, in the case of high powers, some form of cooling. Why the heating effect of electron beams was not exploited in parallel with the applications listed above can be easily explained since heating by other means was more readily achieved. There was also the nuisance of the vacuum that is required to allow a free path to the electrons.

Industrial applications of the thermal effects of an electron beam became feasible only when vacuum technology advanced sufficiently to provide adequate pumping equipment at relatively low cost for the large volumes often required. Now the applications of heating by electron beams cover such a vast and growing area of technology; the following are a few examples:

continuous annealing of sheet metal;
surface treatment, e.g., rapid paint curing;
melting and refining of refractory metals and alloys;
metallization, i.e., the vacuum deposition of very thin metal layers;
fusion welding;
micromachining.

This remarkable versatility is derived from a number of unique and basic features of the thermal effects of electron beams. These are now discussed together with some of the limiting aspects.

Features of heating by an electron beam

Desirable features

CHEMICAL PURITY. The energy of an electron beam is transferred from the emitting cathode to the receiving material through vacuum, unlike an electric arc in which an ionized gas medium is formed to transfer electric energy. Again, the electrons themselves are chemically inert, as they 'disappear' once they are intercepted by solid matter, unlike ions in an ion beam for example. Thus, the combination of purity of the vacuum environment and the inert nature of the electrons leads to chemical purity of processing by electron beams, which is adequate for many of the modern industrial processes demanding a high degree of chemical purity.

GUIDANCE TO SELECTED AREAS. Since the cross-section of the beam can be controlled by focusing, by means of electrostatic or magnetic lenses, and since it can also be readily deflected, it is possible to guide the beam's energy to areas whose size and position are predetermined. The control is simple and the response is rapid since the electron beam has virtually no inertia.

WIDE RANGE OF POWERS. Depending on the application, electron guns have been constructed for thermal applications ranging from a few watts for micromachining operations to furnaces of megawatt capacity for vacuum melting and refining refractory metals and alloys. Even with the same plant, changes in power by a factor of 100 or more can be achieved by simply changing the beam current or the accelerating voltage, or both. For example, the same welding plant can penetrate a steel plate 2in. (50mm) thick and adequately weld a sheet 0·005in. (0·1mm) thick, with no modification to the equipment. With the control which is now available on most equipment, the stability of the beam over such a wide range can be quite adequate.

CONTROLLED POWER DENSITY. The beam power can be spread over an area or focused to a fine spot depending on the degree of focusing. Power densities available from an electron beam vary from 0·1W/mm^2 for curing paint to 10^7W/mm^2 for machining. Indeed, power density is probably the most significant parameter of an electron beam as it controls the magnitude of heating in the irradiated material. The type of thermal application is closely related to power density as illustrated in Fig. 3.1.

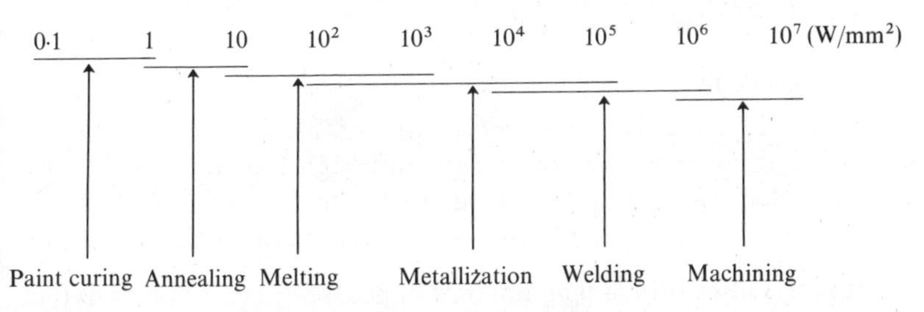

Fig. 3.1 The role of power density in industrial applications

Not only can the overall power density be controlled, but refined electron optics can produce beams whose power density distribution is of a predetermined profile. In welding, it has been claimed that an annular power distribution, rather than the gaussian one normally produced with more conventional units, reduces the incidence of porosity. Indeed, certain commercially available EBW machines incorporate a range of profile-producing electron optics, although it is not clear as yet whether the advantages go beyond the single claim mentioned above.

HIGH POWER EFFICIENCY. Partly because of the absence of a carrying medium and partly because of the mechanism of heating by an electron beam, nearly all the power input to the generating gun is transferred to heat in the receiving material. True, there is the energy 'loss' due to the emission of secondary electrons and the generation of

X-rays, but all this amounts to a small proportion—around one per cent in most cases when a deep weld is being made.

RAPID THERMAL RESPONSE. A variety of reasons contribute to the remarkably rapid response of EB heating. Firstly, the thermal capacity of the emitting heated cathode is generally very small and the high voltage can be switched on quite rapidly. Secondly, no carrying medium is required and the speed of electrons in vacuum is of the same order of magnitude as the speed of light in the range of accelerating voltages of interest here. Thirdly, heating, due to the interception of electrons by solid matter, takes place during the period through which the electrons come to rest, which is of the order of microseconds.

Such rapid response is used to advantage in high-frequency pulsing of the electron beam, thus producing interesting thermal effects, particularly for machining operations.

Undesirable features

There are, naturally, a number of undesirable features associated with heating by an electron beam. Although less numerous than those which are desirable, it is important to take note of them so that, when appropriate, the necessary precautions should be taken.

THE NEED FOR VACUUM. This not only imposes physical limitations on the size of the component to be handled, but has a substantial bearing on the cost of the plant. The vacuum chamber has to be evacuated and the time required has to be included in the manufacturing cycle. Certain materials with a high vapour pressure in the neighbourhood of their melting point will evaporate during fusion processing, thus complicating composition control. Metered additions may have to be progressively introduced to compensate for evaporation losses.

HIGH CAPITAL COST. Reference has already been made to the cost of the vacuum plant. In addition, the cost of the power supply incorporating the high-voltage equipment adds substantially to the total capital outlay. For the same power rating, an EB unit is nearly always more costly than its conventional counterpart.

RADIATION HAZARDS. Whenever an electron beam strikes a metallic target, X-rays are generated. The penetration of X-rays increases with the accelerating voltage of the electron beam. Such hard rays may be capable of penetrating the vacuum chamber walls, in which case lead shielding of the appropriate thickness has to be employed. Manufacturers are fully aware of the international regulations governing the handling of plant with radiation hazards, and machines are designed and constructed well within such regulations. However, any plant alteration brought about by the user must be adequately shielded and checked for radiation leaks.

MAGNETIC FIELDS. Magnetic fields within the vacuum chamber or outside it may deflect the beam from its normal path. This may not be so detrimental in applications that do not demand high positional accuracy, such as paint curing or melting. However, for welding and machining operations, an accurate relative position between the beam and the workpiece has to be maintained. In these cases, the siting of the EB plant must be such that it is not in the immediate vicinity of magnetic field-producing equipment, e.g., resistance welding machines. Again, holding fixtures and other devices placed within the chamber must be non-magnetic. The component itself has to be demagnetized, if this is necessary, although certain materials change their magnetic properties upon thermal treatment during processing. It is possible to compensate for such unavoidable beam deflection by means of the deflection coils in the gun column.

A number of proposals have been put forward, some successfully demonstrated, aimed at 'stiffening' the beam. They all rely on sensing the position of the point of impact of the beam by means of some secondary effect such as the light generated there or the emission of secondary electrons. Any deflection sensed is made to generate a signal which is fed into the deflection coils, bringing about the necessary compensating action.

The energy of an electron beam

Electrons are of exceedingly small mass; $9{\cdot}1 \times 10^{-28}$g. They can, however, be accelerated to very high speeds by the application of a suitable electrical potential. For example, at the readily attainable potential of 150kV, an electron can be accelerated over a very small distance to a velocity of over half the speed of light, or some 100000 miles/s (160000km/s). Kinetic energy is therefore imparted to the electron; the higher the accelerating voltage the greater the energy. Also, the larger the number of electrons in the beam, i.e., the greater the beam current, the greater its energy. Indeed, it is remarkable how many kilowatts of beam power can be derived from quite small filaments.

The kinetic energy of an electron is equal to $\frac{1}{2} \times$ mass $\times$ velocity^2. A single electron accelerated by a potential of one volt will acquire a kinetic energy of one electron volt, where

$$1\text{eV} = 1{\cdot}6 \times 10^{-12}\text{erg}$$

As an example, consider the case of a beam of 10mA accelerated by a potential of 150kV. Since

$$1\text{A} = 6{\cdot}28 \times 10^{18}\ \text{electrons/s}$$

Therefore

$$10\text{mA} = 6{\cdot}28 \times 10^{16}\ \text{electrons/s}$$

Since these electrons are accelerated by 150kV, the beam energy is

$$E = 6{\cdot}28 \times 10^{16} \times 150 \times 10^{3} \times 1{\cdot}6 \times 10^{-12}$$

$$= 1{\cdot}5 \times 10^{10}\text{erg}$$

It is interesting to note that this is equivalent to 1·5kW, since 1W = 10^7erg/s. Thus, the energy of the beam can be derived directly by multiplying together its two basic parameters, beam current and the accelerating voltage. This naturally applies in the case of a steady beam but, for a pulsed or fluctuating beam, the usual corrections will have to be applied.

The range of an energetic electron

When an energetic electron strikes the surface of a solid or liquid substance, it penetrates a thin layer without an appreciable change in its velocity. It thus loses little of its kinetic energy and no heating takes place within this thin layer, which is therefore considered 'transparent' to electrons. The higher the velocity of the electron, the thicker the transparent layer. The electron then encounters resistance from the lattice, causing a progressive decrease in its velocity, until it finally comes to rest. The energy of the impinging electron is transferred to the lattice electrons which, in turn, increase the vibrational energy of the total lattice and heating of the material takes place.

Heating by an electron beam thus takes place internally below the transparent skin, within the range of the electron in the material (Fig. 3.2). The electron range has been studied by many workers and can be expressed

$$\delta = a\frac{A}{Z\rho}V^2$$

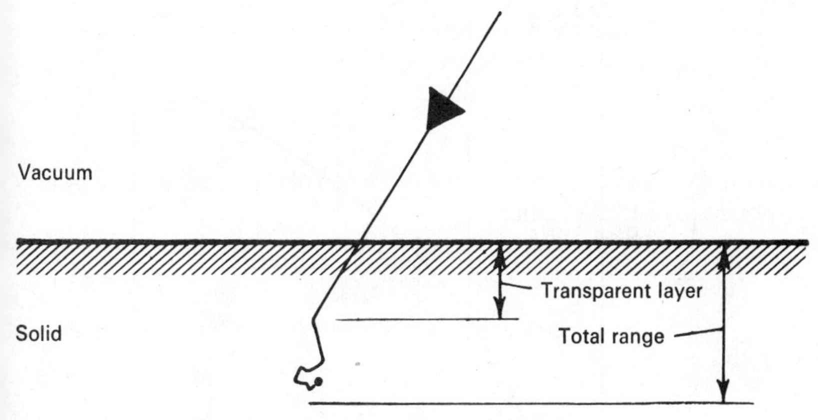

Fig. 3.2 Trajectory of an energetic electron, schematic

where

δ = is the electron range,
a = a constant,
A = the atomic weight of the target material,
Z = the atomic number of the target material,
V = the accelerating voltage of the impingeing electrons, and
ρ = the density of the material.

Thus, the more energetic the electron, the greater the range; but denser materials result in a smaller range for a given accelerating voltage. The above expression is presented graphically[1] in Fig. 3.3, where a scale factor can be applied for each material, taking account of its atomic weight, atomic number, and density. For example, the range of a 100kV electron in iron is nearly 0·001in. (0·025mm). Heating within this layer takes place over the area irradiated by the beam. The extent of heating is dependent on a number of factors.

Naturally, the larger the number of electrons reaching the material surface per unit time, i.e., the greater the beam current, the greater the amount of energy transferred to heat within the material. Thermally conductive materials will dissipate the heat outwards from the irradiated area thus reducing the peak temperature at the centre of the beam. However, heating can become more localized if the beam is pulsed so that, for a short period, before the heat is dissipated by conduction, higher temperatures can be achieved. We shall see that continuous pulsing of an electron beam is utilized in certain applica-

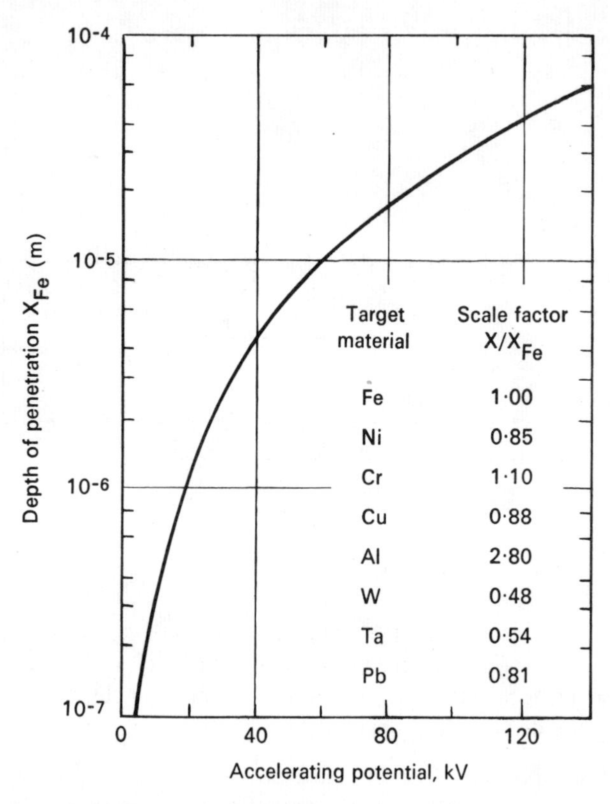

Fig. 3.3 Range of an energetic electron (*Courtesy H. Schwarz*)

tions which benefit from intense local heating. But the most significant factor is the power density of the beam. A well-focused beam of a given power will irradiate a small surface area, thus containing the heating within a small volume. If conduction is not taken into account, it can be readily calculated that temperatures well above the sublimation temperatures of all solid matter can be reached within a fraction of a second. Even when allowance is made for thermal conduction, evaporation temperatures can be reached for most solid materials subjected to a well-focused beam of only a few kilowatts.

The evaporation phenomenon outlined above is the key to deep penetration welding by electron beam. Referring back to the expression for the range of an electron, it can be readily appreciated that

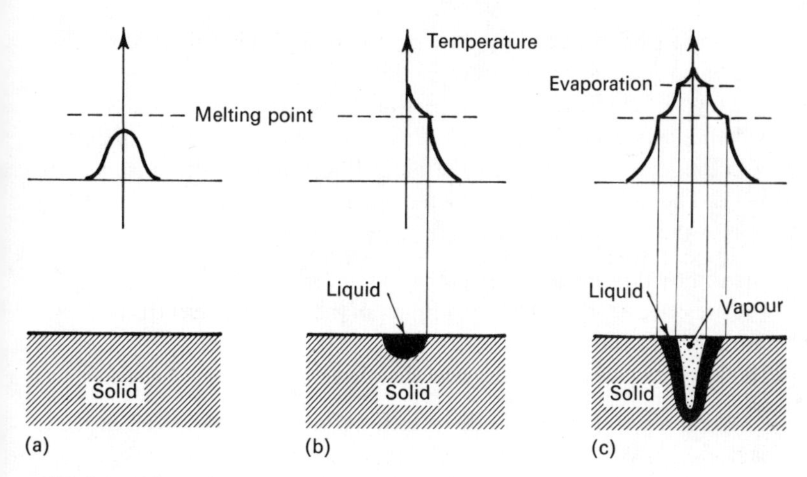

Fig. 3.4 Effect of power density on thermal effects of an electron beam: (a) heating without melting, (b) melting without deep penetration, and (c) deep penetration

the penetration of an electron beam in vapour will be many times greater than that in a solid because of the large decrease in density upon evaporation. Figure 3.4a shows schematically the thermal effect of a low power density beam where no melting takes place. By increasing the power density, higher temperatures are attained at the centre of the irradiated area until melting takes place (Fig. 3.4b). By further increasing the power density, evaporation may occur which leads to a deepening of the molten pool or even the formation of a hole, as in Fig. 3.4c.

Mechanism of deep welding

We have seen that very high temperatures can be attained in the small volume affected by the beam, sufficient in fact to generate metal vapour in the hot zone at the centre of the molten pool. The hole thus produced may disappear because of the contracting forces of the surface tension of the molten ring of material around the vapour centre; alternatively, if the vapour pressure is high enough, the hole will continue to exist as long as adequate energy is fed into the material by the electron beam to maintain the pressure of the vapour. The following analysis attempts to examine the nature of the equilibrium of the forces in the molten pool, thus indicating the temperatures attained in the weld, also giving an insight into some of the basic geometrical characteristics of a deep electron-beam weld.

Meyer, Scheffels, and Steigerwald[2] examined a typical case in which a beam of 130kV and 12mA, focused to a diameter of 0·2mm was found to penetrate a steel plate 10mm thick by a pulse lasting 0·05s. The capillary formed by the beam may have been produced by one of the following processes:

the removal of metal by evaporation;
the pressure produced by the impact of the stream of electrons on the molten material;
the internal pressure of the metal vapour.

We shall now examine the relative contribution of each of these factors.

Capillary formation by simple evaporation

Under the conditions described above, some pyrometric measurements were taken which showed that the temperature at the centre of the molten pool was 2800°K; the rate of evaporation of iron at this temperature is 0·014g/mm^2 s. Thus, for the duration of the beam's pulse of 0·05s, acting over an area 0·2mm in diameter, $2{\cdot}2 \times 10^{-5}$g of metal would be removed by evaporation. However, the mass actually removed to form the capillary is some 120×10^{-5}g. Therefore, direct evaporation can account for only some two per cent of the volume of the capillary.

Electron pressure

When a stream of electrons strikes the surface of a workpiece, a pressure is exerted because the electrons lose momentum. By assuming the electrons lose all their momentum upon impact, the electron pressure p_e can be expressed as

$$p_e = nm_e v$$

where

n = the number of electrons striking the target per unit time per unit area
m_e = the mass of an electron ($9{\cdot}1 \times 10^{-28}$g)
v = velocity of electrons upon impact

The above expression does not take into account the relativistic momentum increase of the electrons, as it is small for the accelerating voltage under consideration.

Now $n = i/e$, where i is the current of the beam and e the charge of the electron. Also, $v = \sqrt{(2eV/m_e)}$, where V is the accelerating voltage. Thus, by substitution,

$$p_e = i\sqrt{\left(\frac{2Vm_e}{e}\right)}$$

From the beam values given earlier, it can be seen that the electron pressure $p_e = 3{\cdot}5$ torr. This pressure will act on the surface of the molten metal producing a cavity whose radius of curvature is

$$r_c = \frac{2\gamma}{p_e}$$

where γ is the surface tension of the molten steel, which, if considered at its melting point, is some 170dyn/mm. Thus, $r_c = 7$mm which is many times greater than the diameter of the capillary, which was found to be approximately the same as that of the beam, namely, 0·2mm. It is clear, therefore, that the electron pressure is weak, producing only a shallow surface cavity.

Pressure of evaporated material

At the measured temperature at the centre of the molten pool of 2800°K, a vapour pressure $p_v = 170$ torr is generated. This is a substantial pressure capable of producing a cavity of a radius

$$r_v = 0{\cdot}14\text{mm}$$

which is of the same order of magnitude as the diameter of both the electron beam and the cavity. It is therefore clear that the metal vapour pressure is primarily responsible for the generation of the capillary.

We shall now proceed to determine the geometrical details of the shape of the capillary taking into account the vapour pressure forces only.

Once a hole is generated by metal evaporation, a molten layer is formed around it. The forces of surface tension will tend to pull outwards from the centre of the hole if the thickness of the material is smaller than its diameter. This is a stable condition which can be applied in cutting by an electron beam or a heat source of comparable intensity. If the thickness of the material is greater than the hole's diameter, however, surface tension will tend to act inwards, thus closing the hole. It is, therefore, more difficult to cut thick materials by means of an intense heat source. If a relative movement exists between the beam and the material, the molten ring will close in after the passage of the beam. This is the situation encountered under welding conditions where the beam cuts its way through the plate while the hole is progressively filled in its wake. If a continuous cutting action is required for a thick plate, a force is needed in addition to the vapour pressure exerted on the molten walls of the cavity. In a plasma jet, this force is provided by the momentum of the jet itself causing expulsion of the molten metal. Such a force is not available from an electron beam and resort has to be made to external methods, such as centrifugal effects or even the vapour pressure of a secondary material placed in contact with the plate to be processed.

If such external forces, which do not operate under welding conditions, are ignored, the shape of the capillary formed by the beam will be decided by the equilibrium of the forces produced by the vapour pressure on the one hand, and the surface tension of the molten metal on the other. The contribution of the weight of the molten column itself can be shown to be too small to make a significant contribution. In deep penetration welding, the surface tension will operate towards the centre of the beam, and this is counteracted by the vapour pressure exerted outwards on the molten walls of the capillary, as shown in Fig. 3.5.

Consider the conditions at the upper regions of the capillary. A bulge is formed above the surface resulting from expulsion of molten metal by the vapour pressure. By equating the volume of the capillary to that of the bulge above the surface, a circular cross-section to the bulge being assumed, and by utilizing the parameters given earlier in this section, it can be readily calculated that the radius of the curvature of the cross-section of the bulge is $r_b = 0{\cdot}2$mm. Considering the equilibrium of the bulge,

$$p_b = \frac{\gamma}{r_b}$$

and for the same value of $\gamma = 170$ dyn/mm, the bulge pressure p_b is then 60 torr.

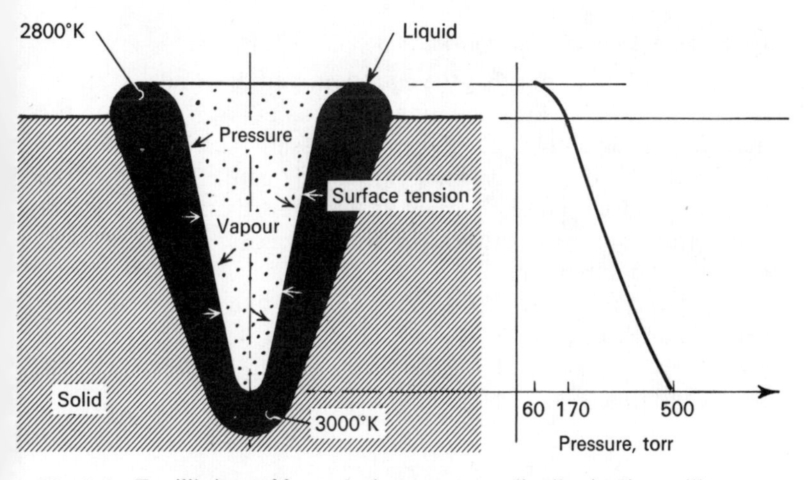

Fig. 3.5 Equilibrium of forces and temperature distribution in capillary

The surface of the molten metal will tilt to form a conical cavity so that the pressure of the surface tension is equal to the vapour pressure exerted on the walls of the capillary. Thus, the radius of the capillary decreases with increasing penetration so that at any point at depth z, the pressure

$$p_z = \frac{\gamma}{r_z} + \frac{\gamma}{r_b}$$

The pressure is highest at the bottom of the capillary where the radius of the hemisphere is smaller than the radius at the capillary opening. The condition of equilibrium at the bottom of the capillary is

$$p_d = \frac{2\gamma}{r_d} + \frac{\gamma}{r_b}$$

where p_d and r_d are the pressure at the bottom of the capillary and the radius of the hemispherical base respectively. From practical observations, r_d is found to be between 0·3 and 0·5 times the radius at the capillary opening. It can be readily calculated that the pressure at the bottom of the capillary is approximately 500 torr. Thus, the vapour pressure rises from 60 torr at the opening to 500 torr at the bottom.

Disadvantages of capillary formation

Meyer *et al.*[3] point out that the equilibrium of the capillary can be maintained only by the continuous addition of beam energy to the material, so that a high enough vapour pressure is maintained. Even so, the equilibrium is unstable, particularly at the bottom of the capillary. Indeed, more rigorous analysis reveals that no state of equilibrium can exist between the surface tension of the spherical cap and the vapour pressure there. Evidence in support of this conclusion can be seen in the irregular form in a weld seam when penetration through the plate is incomplete (Fig. 3.6). Also, shrinkage porosity in deep penetration welds are a practical expression of the same condition (Fig. 3.7).

Fig. 3.6 Irregular penetration in thick plate weld

It is thus argued by the above workers that the deep penetration capabilities of an electron beam are not without certain disadvantages. In addition to the irregularities found in the lower regions of a deep weld, high temperatures are reached there which may have undesirable metallurgical effects such as overheating or shrinkage cracking. There is, therefore, a conflict between the need for sufficiently high temperatures at the root of the weld to maintain a high enough vapour pressure to keep the capillary open, and the detrimental effects of such high temperatures on the material being processed.

Recent developments by Meyer *et al.*[3] were aimed at avoiding the metallurgical disadvantages while at least maintaining the depth of weld under the same specific heat input. One way, for example, would be to relieve the electron beam of the task of opening the capillary, thus reducing the required level of vapour pressure. At the beginning of the seam, a hole is drilled which is slightly smaller in diameter than the electron beam. The beam has mainly to melt a small annular zone surrounding the hole, generating only a poor vapour pressure in the hole. The relative movement between the beam and the plate

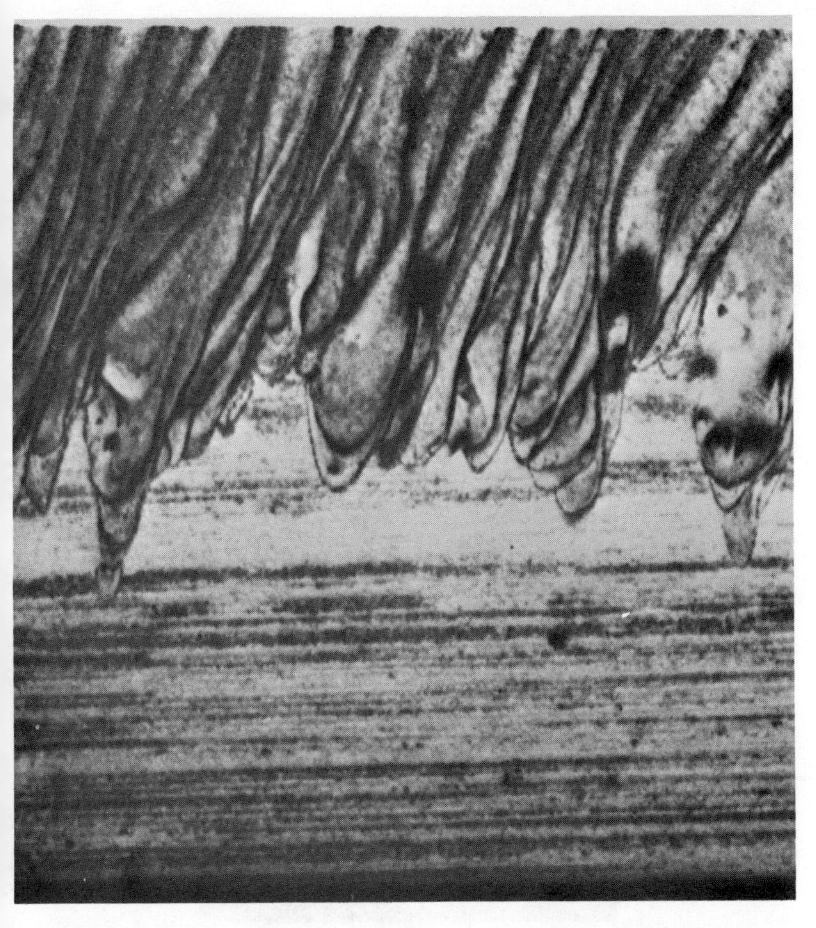

Fig. 3.7 Root porosity in thick plate weld ($\times 30$) (*Courtesy Rolls-Royce Ltd*)

will transport the hole along the seam. The adhesion of the molten ring to the solid material adds to the effect of keeping the capillary open. Thus, with a relatively low beam power density, full penetration is achieved while overheating of the metal structure is avoided, the incidence of spatter at the lower bead, which is normally associated with the expulsion of molten metal by the high vapour pressure in the capillary, is substantially reduced. Figure 3.8 shows the macrosection of an EBW seam produced by the above method.

Fig. 3.8 Deep weld with 'artificial' capillary (*Courtesy K. H. Steigerwald*)

In another development by the same workers, material in the form of wire, ribbon, or granulate is added at the joint line during welding. This method is particularly suited to joining very thick plate, which may be beyond the capability of EBW if conventional operations are used. Wire or ribbon is fed into the narrow gap between the two edges to be welded by equipment not unlike that used in arc-welding operations. However, such equipment has to work satisfactorily in vacuum and has to be capable of more accurate alignment because of the thinner wires used. The deposited material is not directed to a large molten pool as in the case of arc welding; rather it fuses with a narrow surface layer on each side of the two plates to be joined, as shown in Fig. 3.9.

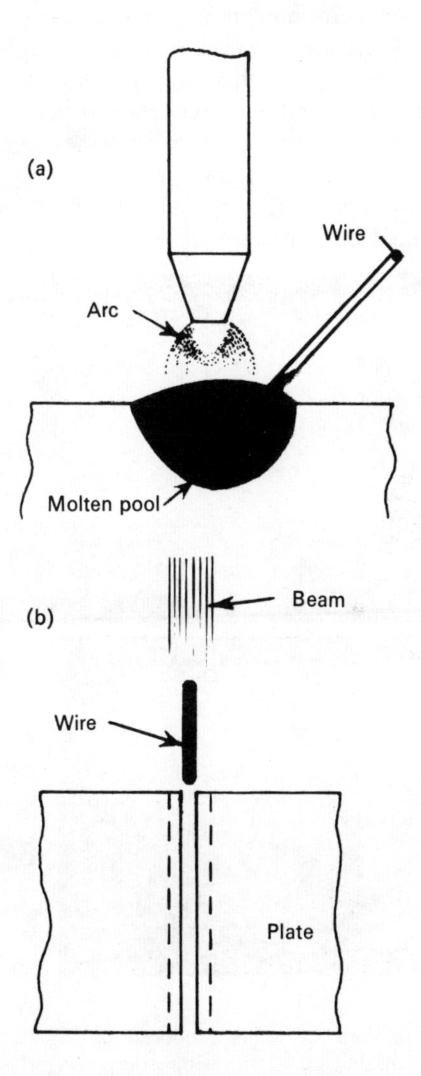

Fig. 3.9 Wire feed used in (a) arc welding, and (b) deep-penetration welding

Comparison of the electron beam with other intense heat sources

Until recently, a well-focused electron beam was the only available heat source of power density of the order of $10^5 W/mm^2$. This monopoly position has now been challenged by the advent of the plasma arc and the laser beam. Both heat sources are capable of performing certain thermal operations, not unlike those produced by an electron beam. There are, however, certain basic differences and it is appropriate to compare the thermal capabilities and limitations of all three sources, with particular reference to welding applications.

It must be pointed out, however, that these techniques are at various stages of development and that some of the present limitations will be overcome. The various relative merits given below must be accepted as a picture of the present position, and the

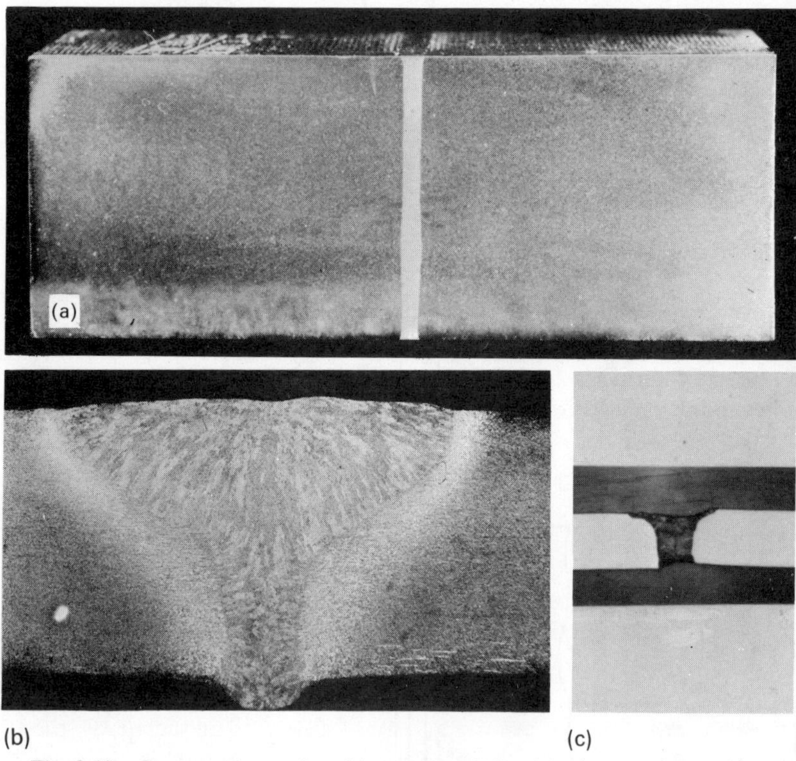

Fig. 3.10 Cross-sections of welds produced by (a) electron beam 1in. (25·4mm) deep, (b) plasma arc 0·25in. (6·3mm) deep, and (c) laser beam 0·040in. (1·016mm) deep (*Courtesy Rolls-Royce Ltd*)

potential user will be advised to follow the progress of each technique as further developments take place.

The similarities of welding capabilities of electron beam, plasma arc, and laser will be considered first. Figure 3.10 shows the cross-sections of welds produced by all three techniques. The similarity of shape is immediately obvious: the weld narrows down with increasing penetration, unlike an arc weld in which the depth-to-width ratio is nearly 1:1, and the ratio is much increased. This indicates a change in the mechanics of fusion from pure conduction for arc welding to some form of cutting action which is responsible for the shape like the stem of a wine glass, or the deep penetration welding regime.

In the plasma arc, intense heat is generated by the ionization of the gas which is fed into the arc. It is possible to produce a weld by means of a plasma arc which is not unlike that of a simple arc, as can be seen in Fig. 3.11a. This can be done by reducing the ionized gas flow through the plasma torch. However, a 'keyholing' regime is established when the momentum of the plasma jet is sufficient to eject the molten metal, and a cutting action is achieved (Fig. 3.11b). Indeed, plasma arc was first introduced as a plate-cutting device.

Thus, the deep penetration capability of a plasma arc is derived from purely mechanical means. There is no evidence that evaporation takes place during plasma welding; indeed, this is not to be expected as the power densities available are not high enough.

The picture is quite different for a laser beam. Power densities are as high as available from a well-focused electron beam and evaporation in the centre of the molten pool is to be expected. Craters are formed and fine holes can be drilled with remarkable similarity to the performance of an electron beam, as can be seen in Fig. 3.12. Also, successful welding operations have been carried out with depth-to-width ratios of up to 5:1. It would appear, therefore, that the mechanism of deep penetration by a laser beam is very similar to that encountered with an electron beam: a crater is formed and the pressure generated by the vapour causes displacement of the molten metal upwards along the walls of the hole.

However, there is one fundamental difference between a laser beam and electron beam. Electrons, although diffracted by the metal vapour atoms, retain a good proportion of their original momentum and are therefore capable of producing further heating right down to the bottom of the capillary. The photons in a laser

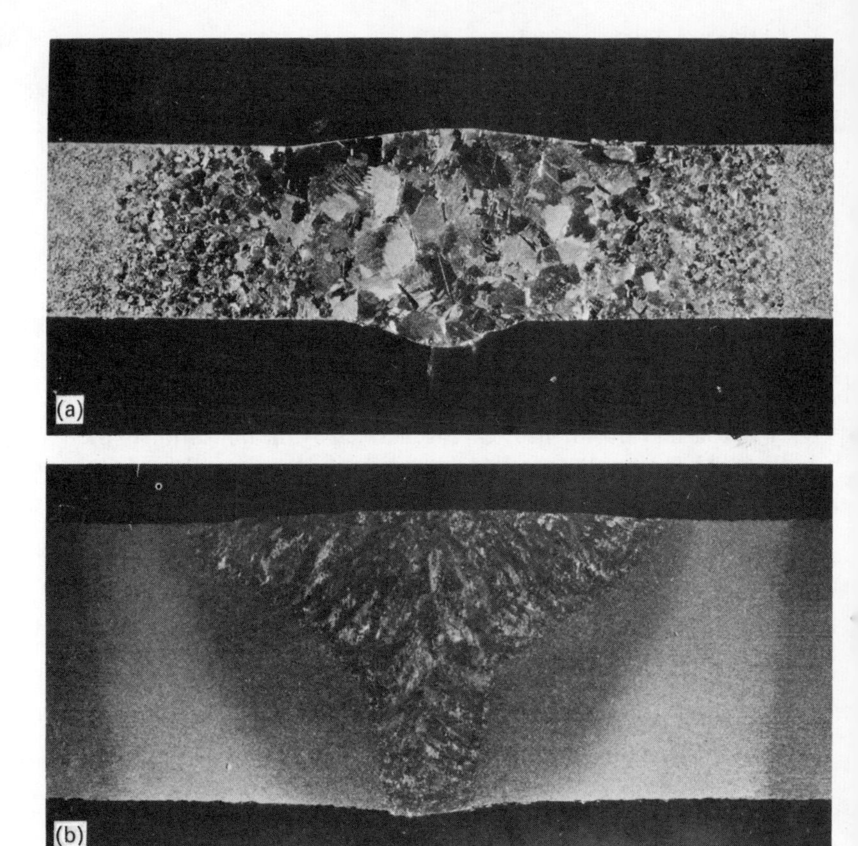

Fig. 3.11 Plasma welds: (a) non-penetrating, and (b) 'keyholing' (×7) (*Courtesy Rolls-Royce Ltd*)

beam have no such capability; stated simply, the metal vapour cloud completely obstructs the beam of laser light. This must impose an upper limit on the penetration capabilities of the beam. This, together with the power limit available from lasers, puts them in a different class of application from an electron beam. While penetration of the order of 2in. (50mm) is normal for an electron beam, the laser normally operates in the region of, say, 0·050in. (1·27mm).

From a practical viewpoint, the electron beam has a feature which is not available from the other two sources. The working distance

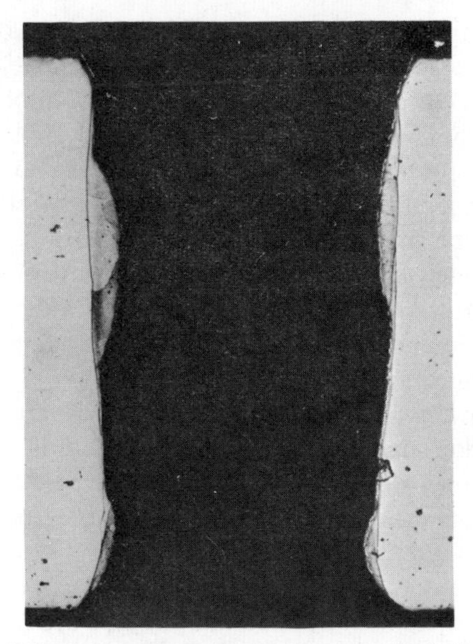

Fig. 3.12 Hole drilled by laser beam (× 90) (*Courtesy Dr B. F. Scott, Birmingham University*)

can be considerably longer, up to 3ft (900mm) or even more; it can also be varied at will be focusing at different levels. This is not so with a plasma arc or a laser beam: the working distances are much smaller—of the order of 1in. (25·4mm)—and cannot be changed readily. The greater flexibility offered by the electron beam is particularly useful when welding components that may otherwise present accessibility problems.

Correlation of welding parameters

The weld penetration of an electron beam is a function of the following parameters:

the beam's own characteristics in the plane of the weld, such as its diameter, convergence angle, current, and accelerating voltage;
the physical properties of the welded material, such as its thermal conductivity and melting point;
the welding speed;
the welding environment.

Many attempts have been made to correlate the above variables, with some success. A universal relationship is unlikely to be found, however, since our present understanding of the mechanism of deep penetration is not complete. There is also the question of variation of beam current profile from one experiment to another, and such variations are even greater from one design of electron gun to another. Even beam diameter is not always recorded, although it is a critical parameter since it decides the power density of the beam and hence its penetration capabilities. Nevertheless, attempts at correlating welding parameters are useful since they may reveal discrepancies due to inadequate theory.

In the following pages, the experimental data are presented and their correlation analysed by reference to certain physical hypotheses. A section is devoted to the effect of the welding environment as revealed from experiments where welding is carried out in a soft vacuum and at atmospheric pressure.

For a given set of welding conditions (acceleration voltage, 140kV; beam current, 4mA; welding power, 0·56kW; 18/8 stainless steel), the weld penetration decreases with increasing welding speed, as can be seen in Fig. 3.13a. This is to be expected, since doubling the welding speed results in a proportionate decrease of energy into the weld material per unit length of weld. However, if such a simple arithmetical relationship controls the process, the penetration should be linearly related to the inverse of the welding speed. The data of Fig. 3.13b are replotted in this manner and it is clear that, at low welding speeds, the penetration falls short of the expected value. This is again understandable, since more heat is conducted through the weld material at low speeds and is thus lost as far as welding is concerned. Another factor is the width of the weld. The greatest penetration for a given beam power is likely to be attained at high speeds since little time is then available for lateral melting. It is the practice to weld at relatively low speeds if wider welds are preferred. Thus, a greater volume of metal is melted at low speeds without contributing to the penetration. This is illustrated diagrammatically in Fig. 3.14.

It was an understandable temptation, therefore, to relate the penetration of a number of alloys to their thermal conductivity.[4] An exponential relationship was indeed found to exist between four alloys (Fig. 3.15) but if the range of the experiment was extended further, immediate discrepancies would have been revealed. For

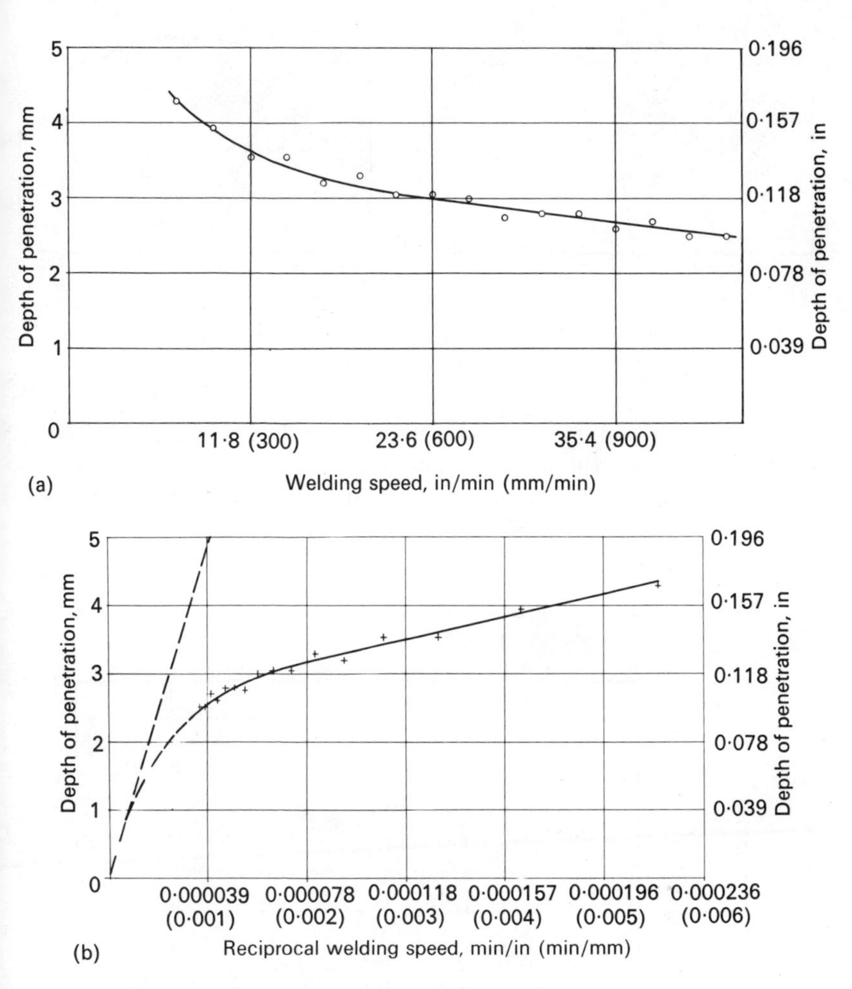

Fig. 3.13 Penetration as a function of welding speed: (a) direct plot, and (b) reciprocal plot

example, although low-alloy steel AISI 4340 has the same thermal conductivity as pure niobium, the penetration in that steel is double that of niobium. It is clear that the discrepancy may be explained by the big difference between the melting points of the two materials.

Hablanian[4] considered both factors, i.e., thermal conductivity and the melting point, and succeeded in correlating the penetration characteristics of a wide range of metals and alloys. It is naturally

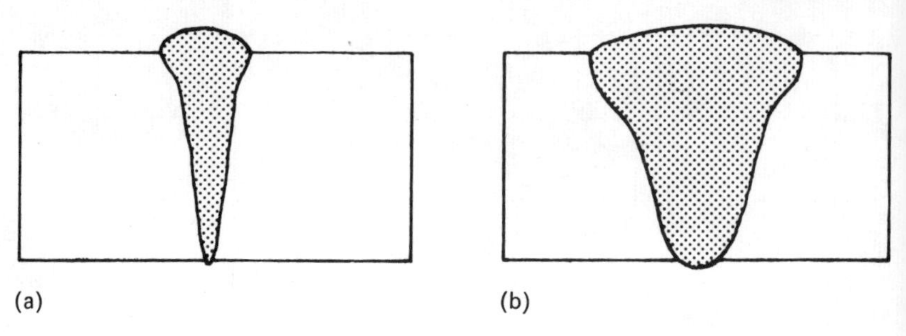

Fig. 3.14 Effect of welding speed on weld width: (a) high speed, and (b) low speed

necessary in such an analysis to reduce the data to dimensionless values, and the resultant plot is shown in Fig. 3.16. The symbols used in the plot are explained in Table 3.1. Since the slope of the straight line is $-\frac{1}{2}$, the relationship can be expressed in the form

$$b = C\frac{P}{Tk}\sqrt{\frac{K}{Vd}}$$

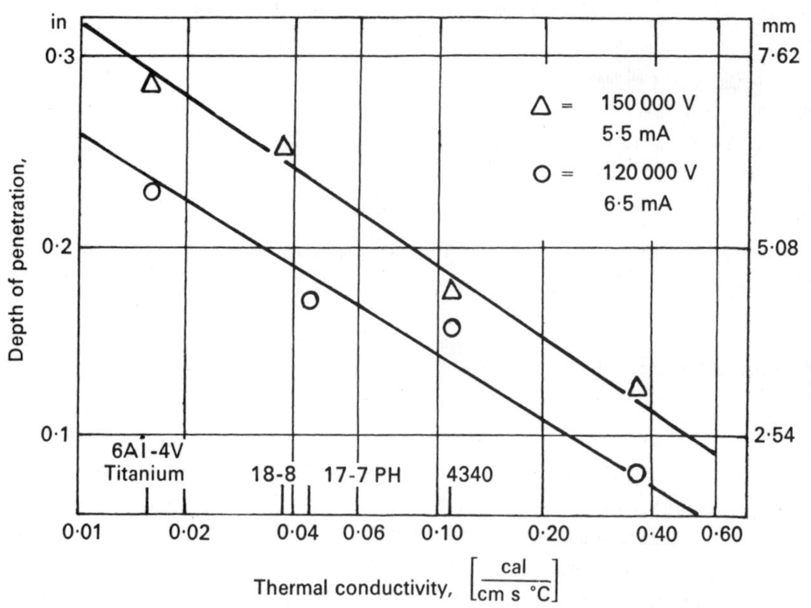

Fig. 3.15 Penetration as a function of thermal conductivity

Thus, the penetration is proportional to the beam power and the square root of the thermal diffusivity. It is also inversely proportional to the melting point, the thermal conductivity, and the square root of the welding speed and the beam diameter. This pattern of behaviour is acceptable in the physical terms used in the analysis; indeed it could have been anticipated from the theory of moving heating sources.

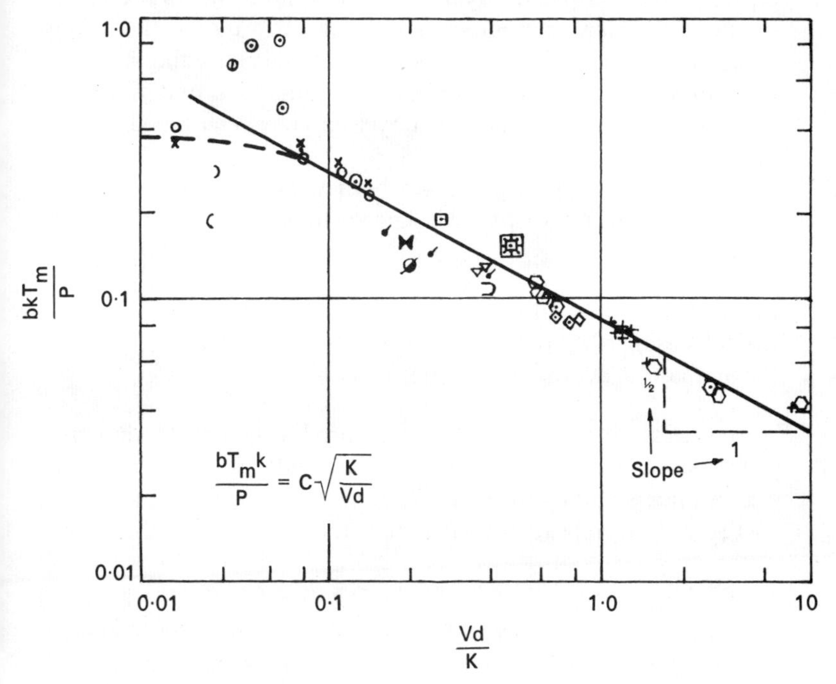

Fig. 3.16 Penetration as a function of thermal conductivity and melting point (*Courtesy M. H. Hablanian*)

Such a tidy pattern would have been completely acceptable but for the irregular behaviour of aluminium and beryllium: aluminium exhibits penetration lower than anticipated but beryllium shows greater penetration. Hablanian explains the latter discrepancy by the unusually low melting point of beryllium, but is unable to find an acceptable explanation for the deviation of aluminium. However, he rightly accepts that his analysis is incomplete as it does not take account of the mechanism of cavity formation.

As described earlier in this chapter, the mechanism of deep penetration is closely related to the evaporation of the welded material. Passoja[5] exposed a clear relationship between penetration and the standard heat of formation of vapour as shown in Table 3.2. He carried out a dimensional analysis taking into account the thermal conductivity of the material, its melting point, and the heat of formation of vapour. The close correlation between theory (smooth curves) and experimental data is clearly seen in Fig. 3.17. Passoja constructed a number of curves correlating penetration with welding speed and beam current for a range of materials including magnesium, aluminium, copper, stainless steel, tantalum, and tungsten. Their validity has yet to be tested by accurately determined experimental data.

Even then, no analysis has yet been worked out to take into account the variables which are known to affect penetration such as:

the surface tension of the molten metal;
the defocusing effect of the vapour by electron scattering;
the mechanical action of the vapour jet.

Nevertheless, theoretical analysis is producing results that are progressively converging towards experimental data, indicating a clearer understanding of the mechanism of deep penetration of an energetic electron beam into solid materials.

The effect of the welding environment on weld penetration

Until very recently, nearly all EBW was carried out in a vacuum of 10^{-4}–10^{-5} torr. This was probably because such a degree of vacuum was required to generate the beam and it was convenient in the early days of EBW to maintain the same level of vacuum in both the electron column and the working chamber. With low-voltage guns placed inside the chamber, there is obviously no choice. Again, most early applications of EBW involved smaller parts with correspondingly small chambers that could be readily evacuated to a 'hard' vacuum. Now that the technique has found increasing application in general engineering, larger chambers have had to be built to accommodate large parts. It soon became obvious that there was economic merit in welding at higher pressures, thus reducing the cost of the pumping plant. However, it was yet to be resolved

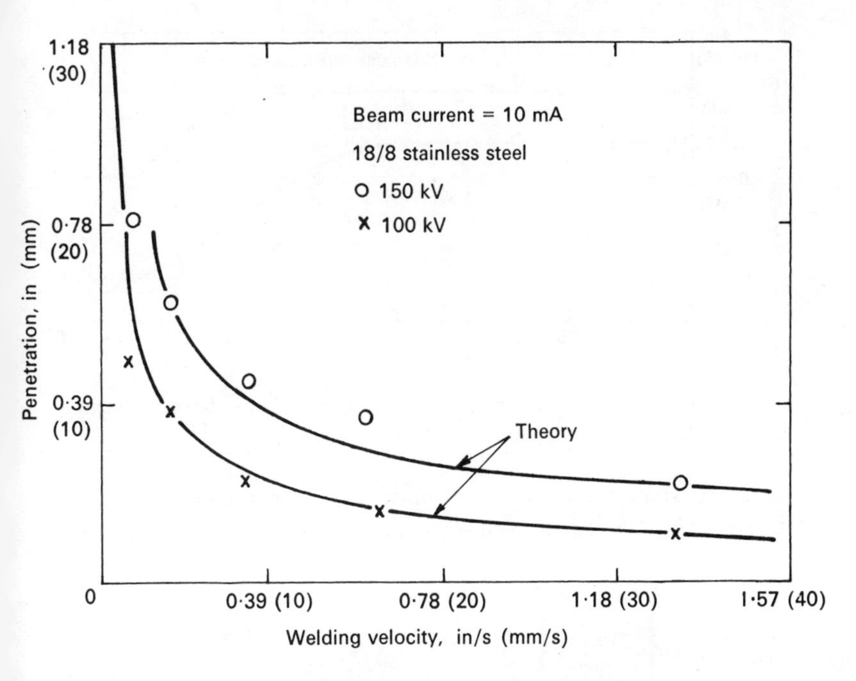

Fig. 3.17 Penetration as a function of thermal conductivity, melting point, and heat of formation of vapour.

whether the more dense environment of a 'soft' vacuum would have an adverse effect on the penetration capabilities of the beam.

It is clear that the presence of gas molecules in the path of the beam would result in some loss of beam power by absorption, and a degree of electron scattering. Scattering is, in effect, a defocusing action that would reduce the power density of the beam. Naturally, this is a matter of degree, and experimental work was undertaken, for example by Meier,[6] to establish how far the working pressure could be allowed to increase before significantly reducing the beam's penetration. Figure 3.18 shows the effect of air pressure in the chamber on weld penetration when welding AISI 4340 with a beam voltage of 100kV and current of 9mA. Data are presented for two working distances, 1in. and 10in. (25·4 and 254mm). The working distance is measured between the heat shield and the point of impact of the beam on the testpiece surface. The beam path in the higher pressure environment is 15in. (380mm) longer than the working

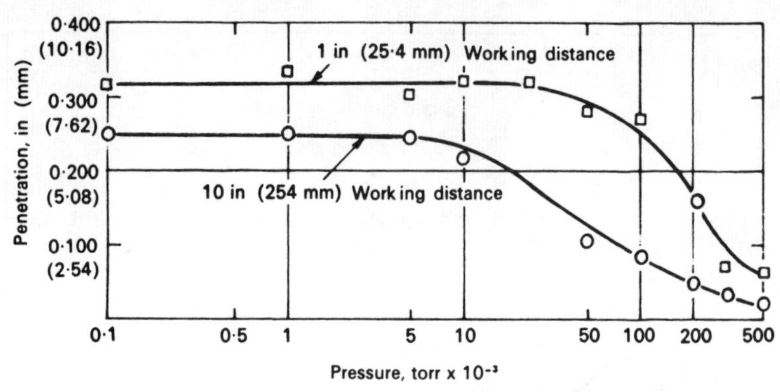

Fig. 3.18 Effect of air pressure on weld penetration (Note: beam path length at indicated pressure is equal to working distance plus 15in. [381mm])

distance, since the lower position of the electron column is maintained at the same soft vacuum as the working chamber (Fig. 3.19).

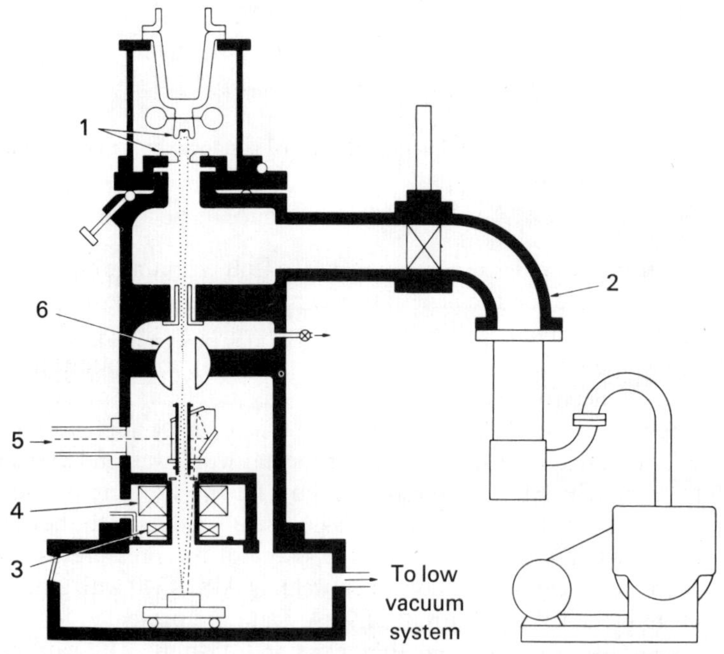

Fig. 3.19 Cross-section of soft-vacuum column: 1—electron-beam gun; 2—high-vacuum pumping module; 3—deflection coil; 4—magnetic lens; 5—optical viewing system; 6—column valve (*Courtesy Hamilton Standard Division of United Aircraft Corporation*)

It is clear from the data shown in Fig. 3.18 that no significant reduction in penetration takes place until the chamber pressure exceeds 2×10^{-2} torr for a working distance of 10in. (254mm); for the 1in. (25·4mm) working distance, the 'critical' pressure is even greater. Thus, the pressure can be increased some two hundred fold without an adverse effect on penetration. This is a significant observation demonstrating the feasibility of 'soft' vacuum welding.

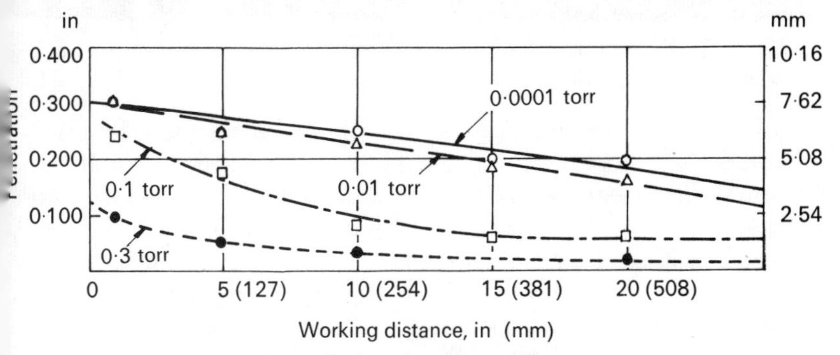

Fig. 3.20 Effect of working distance on penetration

The effect of the working distance is significant: the longer the path of the beam in the more dense environment, the greater the beam spread. The effect of the working distance on penetration for a range of chamber pressures is shown in Fig. 3.20 for the same basic parameters as those relating to Fig. 3.18 It is significant that, for example, a penetration of 0·2in. (5mm) can be achieved at a working distance of 15in. (380mm) and a pressure of 10^{-2} torr but that when the pressure was increased to 10^{-1} torr, the working distance had to be decreased to 3in. (75·2mm) to achieve the same penetration. Thus, the shorter the working distance, the smaller the effect of increased pressure.

The fusion-zone geometry is also affected by the pressure in the chamber. Figure 3.21 shows that no significant modification takes place when the pressure is increased from $0{\cdot}1 \times 10^{-3}$ up to 50×10^{-3} torr. Progressively, however, a decrease in depth-to-width ratio is noted, and at a pressure of 300×10^{-3} torr the deep-penetration capabilities of the beam are nearly lost. It is important to note that a similar series of illustrations could be produced by progressively defocusing the beam or by increasing the working distance.

The same trend continues when the pressure is increased to

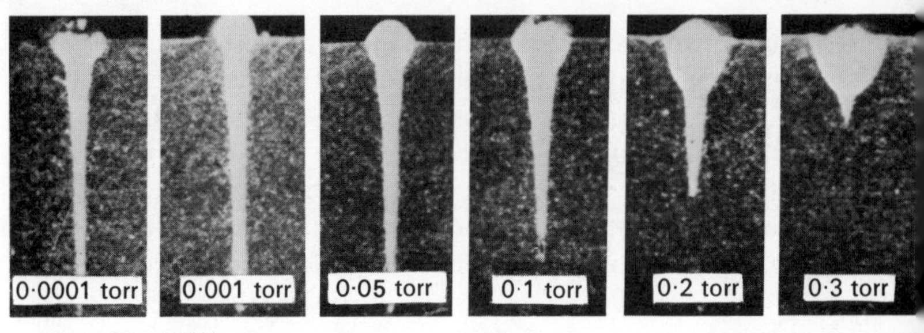

Fig. 3.21 Effect of gas pressure on weld geometry (*Courtesy Hawker Siddeley Dynamics, Electron Beam Division*)

atmospheric. However, the loss of beam power by absorption becomes much more significant and it is therefore necessary to utilize a beam of a higher accelerating voltage and to bring the workpiece as close to the orifice as possible. Typical values for beam voltage when welding in air are 150–200kV and the working distance is seldom greater than $\frac{3}{4}$in. (18·8mm). Although deep penetrations can be achieved, the weld zone becomes wider than in welding under vacuum conditions. This is illustrated in Fig. 3.22 which shows a penetration in stainless steel of $\frac{3}{4}$in. (18·8mm) which was achieved by the application of a powerful beam of 12kW. The effect of working

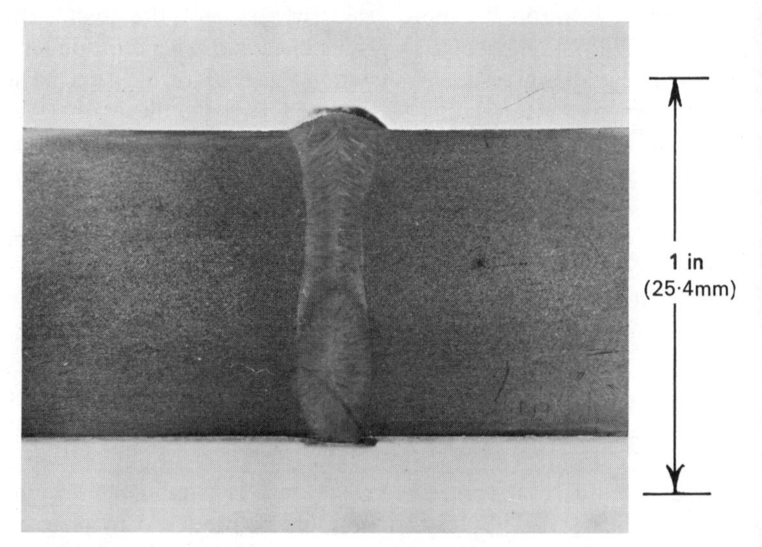

Fig. 3.22 Deep weld produce in air (*Courtesy Hawker Siddeley Dynamics, Electron Beam Division*)

Table 3.1 Symbols used in plotting Fig. 3.16 (*Courtesy M. H. Hablanian*)

Symbol	Material	Power (W)	Speed (in/min)	Speed (mm/min)	Reference*
⊃	Stainless steel	2300	6·3	160	3
⊖	Aluminium 2024	1450	10	254	4
⊡	Stainless steel	1450	15	381	4
⊙	Beryllium	260	15	381	5
	Beryllium	275	15	381	5
	Beryllium	450	15	381	5
	Beryllium	840	80	2032	5
●	AISI 4340	1950	27	686	1
○	AISI 4340	1250	27	686	1
×	AISI 4340	700	27	686	1
∅	Titanium	825	27	686	1
△	Titanium	780	27	686	1
⊡	Stainless steel, 18/8	1350	27	686	1
	Stainless steel, 18/8	1450	27	686	1
D	Aluminium	825	27	686	1
ɑ	Aluminium	780	27	686	1
⬭	Stainless steel, 18/8	4750	50	1270	NRC
⬭ (dotted)	Stainless steel, 18/8	4750	25	635	NRC
+	Stainless steel, 18/8	4750	15	381	NRC
◇	Stainless steel, 18/8	4500	8	203	NRC
∅	Aluminium	3350	24	610	NRC
⋈	Aluminium	2100	24	610	NRC
●́	Zirconium	400	12	305	NRC
	Zirconium	3300	12	305	NRC
	Zirconium	4500	12	305	NRC
◬	Stainless steel	5000	15·5	394	NRC
◈	Titanium	2000	17	432	NRC
	Titanium	2300	17	432	NRC

* References relate to those given in Hablanian's paper.[4]

Table 3.2 Relationship between penetration and standard heat of formation of vapour

Material	Standard heat of formation of vapour (kcal/mole)	
Tungsten	201	
Tantalum	185	↓
Niobium	184·5	
Molybdenum	155·5	Increasing
Nickel	101·61	penetration
Iron	96·68	
Copper	81·5	↓
Chromium	80·5	
Aluminium	75·0	
Magnesium	35·9	

Table 3.3

Material	Thickness (in)	Thickness (mm)	Voltage (kV)	Current (mA)	Welding speed (in/min)	Welding speed (mm/min)	Working distance (in.)	Working distance (mm)	Atmosphere
Killed steel	0·125	3·17	175	30	85	2159	$\frac{1}{4}$	6·35	air
SAE 1010 steel	0·010	0·25	150	2·5	145	3683	$\frac{1}{4}$	6·35	air
SAE 4620 steel	0·400	10·16	175	40	26	660	$\frac{3}{16}$	4·76	air
AISI 4130 steel	0·025	0·63	175	4	80	2032	$\frac{1}{4}$	6·35	4He-1A
AISI 4340 steel	0·250	6·35	175	40	60	1524	$\frac{1}{2}$	12·7	3He-1A
AISI 4340 steel	0·430	10·92	175	40	15	381	$\frac{3}{16}$	4·76	4He-1A
AISI 321 stainless steel	0·060	1·52	175	20·5	130	3302	$\frac{1}{4}$	6·35	air
AISI 304 stainless steel	0·280	7·11	175	29	36	914	$\frac{1}{4}$	6·35	air
AISI 304 stainless steel	0·400	10·16	170	35	10	254	$\frac{1}{4}$	6·35	air
René 41 high-temperature alloy	0·225	5·71	175	40	80	2032	$\frac{1}{2}$	12·7	4He-1A
René 41 high-temperature alloy	0·275	6·98	175	40	15	381	1	25·4	4He-1A
Inconel X	0·125	3·17	150	25	18	457	$\frac{3}{16}$	4·76	air
TZM molybdenum	0·187	4·75	175	40	15	381	$\frac{1}{2}$	12·7	He
5052 aluminium alloy	0·0093	0·236	150	3	145	3683	$\frac{1}{4}$	6·35	air
2219 aluminium alloy	0·375	9·52	175	40	55	1397	$\frac{3}{8}$	9·52	He
Copper	0·008	0·20	150	8	145	3683	$\frac{1}{4}$	6·35	air

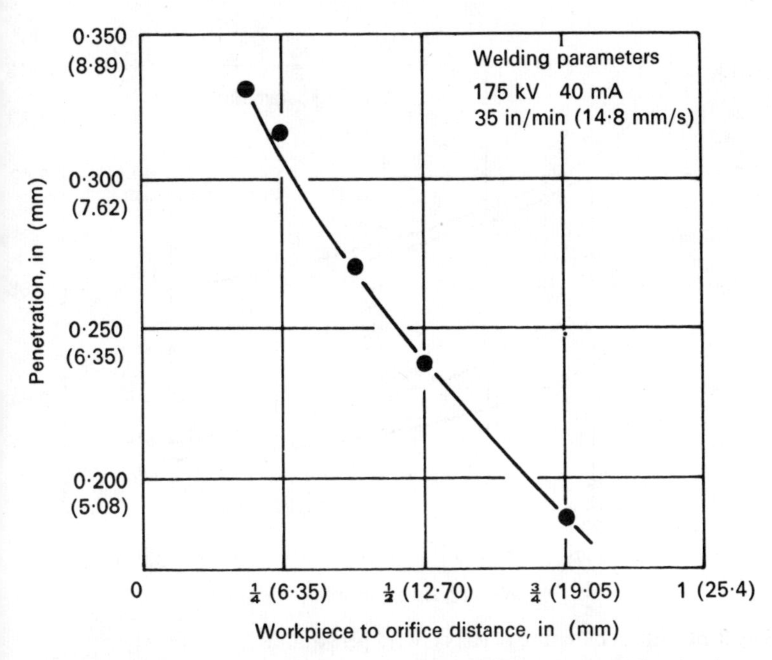

Fig. 3.23 Effect of working distance on penetration when welding in air

distance on penetration is shown to be significant in Fig. 3.23. Under the welding conditions listed, the penetration falls rapidly with increasing working distance.

The extent of scattering and absorption of electrons in a gaseous environment is a function of the structure and mass of the gas molecules. Thus, it is to be expected that the nature of the gas will have some effect on the penetration characteristics of an electron beam. There is also the related chemical effect, which may be significant in the welding of reactive materials where it may be advisable to admit into the chamber some inert gas such as helium, rather than to use air which may cause some contamination or oxidation.

Meier[7] examined the effect of the nature of the gas at atmospheric pressure on the penetration of AISI 4340 steel. The results shown in Fig. 3.24 illustrate that the dense argon atmosphere decreases the penetration by a factor of up to four as compared with the lighter helium. Typical non-vacuum welding parameters for a range of materials and atmospheres are listed in Table 3.3.

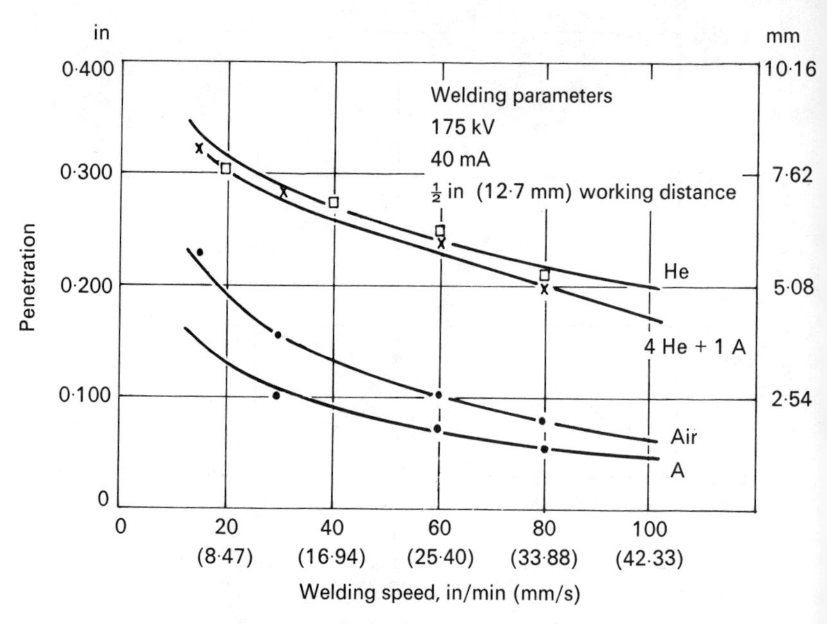

Fig. 3.24 Effect of nature of gas on weld penetration

Metallographic examination and mechanical tests were undertaken by Meier on welds produced at atmospheric pressure. These showed that good welds had been produced, thus demonstrating that, at least for the range of materials examined, welding at atmospheric pressure has no detrimental effects on weld quality.

REFERENCES

1. SCHWARZ, H. 'Mechanism of high power density electron beam penetration in metals.' *J. App. Physics*, **35**, 1964, 2020–9.
2. MEYER, W. E. *et al.* 'A study of the formation and the energy balance of the capillary in electron beam deep penetration welding.' *Procs of the Electron and Laser Beam Symposium*, Pennsylvania State University, 1965, 531–40.
3. MEYER, W. E. *et al.* Private communication.
4. HABLANIAN, M. H. 'A correlation of welding variables.' *Procs of the Electron Beam Symposium*, 5th Annual Meeting, Boston, Mass., 1963, 262–8.
5. PASSOJA, D. E. 'Penetration of solids by high-power-density electron beams.' *Brit. Weld. J.*, **14** (1), 1967, 13–16.

6. Meier, J. W. 'Electron beam welding at various pressures.' 2nd International Conference on Electron and Ion Beam Science and Technology, New York, 1966.
7. Meier, J. W. 'Recent developments in non-vacuum electron-beam welding.' 1st International Conference on Electron and Ion Beam Science and Technology, Toronto, 1964. Wiley & Co., New York, 1965, 634–54.

4. Equipment

Previous chapters have already described how the EBW process has evolved, the fundamental basis of beam generation and control, and the thermal effects of a well-focused beam when it impinges on solid material. The exciting possibilities of how the beam may be employed will have been appreciated by the reader, but, for these possibilities to be realized in practice and on an industrial scale, equipment capable of effective beam generation and control must be made commercially available. It is possible to construct an electron gun and to generate a beam by means of simple devices—many research workers have done so over the years—but there is a long step between an experimental piece of equipment and an industrial plant. Probably the most significant difference is in the degree of control. Control covers far more than focusing the beam or its deflection; there is the control of beam power and the beam's position relative to the workpiece, and also the stability of such parameters over extended periods of welding time.

The electron beam is only part, albeit the essential one, of the integrated complex of an EBW plant. This incorporates the vacuum chamber and pumps, work-handling mechanisms, optical viewing systems, and operator controls. The success of the welding operation will depend on the satisfactory functioning of all these components, and many manufacturers take the view that the equipment should be supplied as a complete system with automatic sequential pumping and interlocking safety circuits. This is an essentially sound philosophy.

The choice of a machine system is usually governed by the nature of the application and the quantities involved. A machine system may be required to weld a wide range of products in a range of sizes, but the quantities involved may be small in each case. This may be the nature of the work load in a jobbing shop. The machine may also be required to weld large quantities of the same component, still with a large diversity in the range of products. Mass production demands special machine features: the size of the component may be so large as to preclude the use of a work chamber, or it may be so small that high-powered optical viewing systems are necessary. The materials may be highly reactive and require a good vacuum to avoid contamination. The potential user of EBW is fortunate in that, in spite of the short history of the process, electron-beam equipment is commercially available to meet almost any of the above requirements.

There are many ways in which commercially available equipment may be classified. Irrespective of type, the electron gun is housed in a vacuum better than $0{\cdot}5 \times 10^{-3}$ torr, as an electron beam cannot be created and maintained unless the vacuum is of this order, but the work area need not be maintained at the same level of vacuum. Depending on the environment surrounding the workpiece, equipment can be divided into the following three categories.

Hard-vacuum environment

In these machines the work chamber is evacuated to the same level as the gun environment, i.e., better than $0{\cdot}5 \times 10^{-3}$ torr; pump-down times vary according to chamber size and pump capacity. The majority of the systems developed to date and commercially available operate under these conditions.

Soft-vacuum environment

In addition to mechanical pumps and/or Rootes pumps, oil diffusion pumps are required to evacuate chambers down to a level of 10^{-4} torr, and with large chambers high costs are involved if short pump-down times are required. A vacuum of 10^{-1} torr can be created without a diffusion pump and in much less time, thus reducing both capital cost and the overall weld cycle time. These environments are used in large chambers and also in chambers which are tailor-made to accommodate only one type of workpiece and its attendant work-handling devices. Such machines are designed to be simple and rapid in operation to permit high production rates. Single- and multi-station units are available with quickly detachable small work chambers and fixtures to facilitate production of a variety of parts.

Non-vacuum environment

The electron beam is brought into the atmosphere by using a specially designed transition chamber module. This chamber reduces the differential between the 10^{-4} torr pressure of the gun to atmospheric pressure by permitting the beam to pass through a series of narrow chambers each evacuated to a progressively higher pressure. The workpiece is fixtured in air but shrouding can be employed if required. The elimination of the chamber opens up the field of applications to include any size of component. There are, however, certain disadvantages in non-vacuum EBW which will be discussed later in this chapter.

The selection of the most appropriate type of machine is dependent on the nature of the materials to be welded: component geometry and size, component dimensional tolerances, the quantities involved, and the cost. The high production rate capabilities of the soft-vacuum equipment are naturally attractive to the mass-production applications required in, for example, the automobile industry, and the equipment is also simple to operate. The hard-vacuum systems provide the purest environment and the range of such machines commercially available offers general-purpose versatility.

Non-vacuum equipment, however, has to be considered more closely. One of the main advantages of EBW is its ability to weld with the minimum of heat, thus causing little thermal distortion and the minimum of metallurgical disturbance to the material. The full vacuum environment results in the highest power density and hence the lowest heat input. As the pressure rises, the power density decreases and the amount of heat entering the workpiece increases, so that the distortion level is greatest with the non-vacuum electron beam. Nevertheless, it is still less than experienced with other fusion welding processes.

The main purpose of the following pages is to provide the potential user of EBW with an insight into the basic features of commercially available equipment so that he may make a wise selection. It would be futile simply to recite the documentation of such equipment; the potential user can acquire such data direct from the manufacturers. In any case, in a book of this nature, it is more useful to deal with basic features than to describe details of current equipment which may conceivably become out of date in a young and developing technology.

Commercially available equipment

It will be appreciated from the chapter on History that EBW systems have been developed independently in several countries. Although the first systems of any commercial significance were developed in Europe, companies in the United States have acquired ownership of patents and have successfully developed production welding machines. Table 4.1 lists the companies known to produce equipment. Although the bulk of equipment so far delivered comes from two sources, namely Hamilton Standard and their associates, and

Table 4.1 Some known suppliers of equipment

I *In the United Kingdom*
Bir-Vac Limited
British Oxygen Company Limited (Edwards High Vacuum Limited)
Cambridge Vacuum Engineering Company Limited (Licensees of Brad Thompson Industries Incorporated)
Hawker Siddeley Dynamics Limited (Licensees of Hamilton Standard)
Sciaky Electric Welding Machines Limited (Licensees of Sciaky Brothers Incorporated)
Vickers Limited
Wentgate Engineers Limited

II *In European Countries*
'Alcatel' Materials Enervide—France
Leybold-Heraues GmbH and Company—W. Germany (Licensees of Hamilton Standard)
Précis (S.A.B.)—France
'Sciaky'—France
S.E.A.V.O.M. (Société d'Etudes et d'Application de Vide Optique et Mécanique) —France
Steigerwald Strahltechnik GmbH—W. Germany

III *In the United States*
Alloyd General Corporation
Bendix Balzers Vacuum Incorporated
Brad Thompson Industries Incorporated
Electro Glass Incorporated
Hamilton Standard Division of United Aircraft Corporation
National Research Corporation
Sciaky Brothers Incorporated
Westinghouse Electric Corporation
Thomson Electric Welding Company Incorporated

IV *In Japan*
Jeol Products (Japan Electron Optics Laboratory Company Limited)
Nippon Electric Company Limited (Licensees of Hamilton Standard)

This list is a guide only and does not claim to be comprehensive.

Sciaky, the potential user is advised to examine the range of products available from all sources. Factors to be taken into account should include not only the considerations outlined above, but also considerations such as supporting technical services and development and supply of accessories.

The information given in Table 4.1 cannot be claimed to be comprehensive; in any case the picture is likely to change with the growing demands of industry.

The work chamber

Vacuum chambers for EBW have been made in many shapes and sizes. Manufacturers normally supply a standard range but are naturally willing to construct special chambers to suit special requirements. The size of the chamber has a marked influence on the total cost of the plant, not only because of the cost of manufacture but also because of the size and capacity of the pumping system. The chamber selected should therefore be no larger than is necessary to accommodate the workpiece and the associated manipulating fixtures.

Chambers are normally constructed from mild steel or stainless steel. If mild steel is used, some surface treatment, such as tinning, is required to combat corrosion. However, mild steel is particularly attractive since it shields the electron beam from stray magnetic fields which may be generated outside the vacuum chamber, but the outgassing of a mild-steel surface is three times slower than that of stainless steel. There is also the obvious cost disparity. Mild steel clad with stainless steel is employed with advantage over solid stainless steel and is probably the most acceptable compromise. (Epoxy-coated mild steel has also been successfully employed.)

The steel plate is normally welded to a high standard and the welds are tested to reveal any leakage. The chamber, as a structure, is designed to withstand the forces of atmospheric pressure without undue deflection. The doors are either hinged or slide on rails to allow access to the chamber interior. Doors normally actuate limit switches which ensure that the high-voltage circuit to the gun is interrupted when any door is open.

There are two general shapes of vacuum chamber: rectangular or cylindrical. The cylindrical form is inherently more rigid and is favoured for large chambers. The rectangular construction allows for chamber extensions to be added to any of the side walls of the chamber. Generally, however, the nature of the application decides the most appropriate chamber shape.

The question of X-radiation requires special attention. A small proportion of the electron-beam energy is converted to X-rays which must be intercepted by the chamber walls. If the accelerating voltage of the electron is below, say, 60kV, thick steel walls will be sufficient to absorb the X-rays; at higher voltages, lead shielding must be added. The EBW plant manufacturers are fully alive to this aspect of chamber design and provide all the necessary screening. Tests are carried out to ensure that from no part of the machine is the leakage rate above the limits determined by internationally recognized regulations for non-supervised installations. In the United Kingdom, the maximum allowable level is 0·75mrad/h.

In addition to the illumination provided for the optical viewing system, some general illumination is required in the chamber. This is useful for setting-up purposes before the final adjustments. The light should be powerful and the bulb placed in such a position that it is not unduly exposed to the metal vapours generated during welding.

Viewing ports are normally provided. The absence of metal there demands that the glass used should be both mechanically strong to withstand atmospheric pressure and capable of absorbing X-rays. The windows are normally composites of 1in. (25mm) thick annealed glass for strength, with a lead glass layer of $\frac{1}{4}$–$\frac{1}{2}$in. (6–12mm) for X-ray absorption. Inner and outer hard-glass panes are provided; the inner one to protect the annealed glass from metal vapour deposition and the outer pane to provide general mechanical protection.

Additional spare ports with blanking plates are provided for two main purposes. It is often necessary to place the gun at some extreme position rather than in the geometrical centre of the chamber—for example, when performing peripheral weld of a large diameter. The ports in the ceiling of the chamber are provided for such an event. If the gun is supported on a sliding seal arrangement, the need for additional ports for this purpose is naturally not as great. The second reason for providing additional ports is to allow access to ancillary equipment such as the wire-feed drive mechanism or an additional manipulator which may be more conveniently placed outside the vacuum chamber. In such cases, the ancillary unit is housed in a box with the necessary X-ray shielding, and the assembly is bolted to the chamber against the port.

It is often convenient to introduce only the drive shaft of an ancillary piece of equipment into the chamber. The rotary seal ports used in this case are provided with double O-ring seals and the interspace between them is held under vacuum by a separate line from the holding pump. A special arrangement is used in the design of the ports to prevent X-rays escaping through the joints.

Power is often required to operate certain accessories placed in the chamber such as additional lighting, electric motors, heating devices, and thermocouples. A generous number of terminal connections—up to twenty—is generally provided for such requirements.

Handling work

The selection of the work-handling system is closely related to the nature of the part to be processed by the machine. Consider, for example, the case where only saw blades are produced. All that is required is a work-handling device capable of linear movement. If the saw blades are produced in a continuous manner, a two-roll system would be adequate. Again, if only round capsules are to be produced, a rotary work-handling system would suffice, although

Fig. 4.1 Early Zeiss machine with universal, five-axis, manipulator (*Courtesy Hamilton Standard Division of United Aircraft Corporation*)

Fig. 4.2 Mobile Sciaky gun within vacuum chamber (*Courtesy Sciaky Electric Welding Machines Ltd*)

facilities must be provided for performing the rotational movement about any axis. Some mode of tilting the drive mechanism should therefore be provided.

When considering the other extreme case, that of a jobbing shop, the diversity of the tasks may dictate the selection of a universal work-handling system. In the early days of EBW, such universal systems were generally provided (Fig. 4.1), since the manufacturer was unable to determine the exact requirements of his potential customers. The general trend is bound to be in the direction of selecting special-purpose work-handling systems to meet well-specified requirements.

Most standard machines are supplied with a number of basic work-handling devices, which can be augmented when required. The worktable is power-driven to provide movement in the horizontal plane in two orthogonal directions, X and Y. This is the basic arrangement for a fixed gun. In cases where the gun is mobile within the chamber, the X motion along the chamber axis is provided by the work-handling device, while the Y motion is supplied by the gun's own traversing mechanism (see Fig. 4.2).

Movement in the vertical or Z direction is provided either as a positional adjustment of the height of the workpiece or as a continuous drive movement suitable for welding. In the latter case, such a movement is useful when welding a component whose surface is inclined at an angle to the horizontal; a combined X and Z movement at the appropriate speeds then maintains the surface of the workpiece at a constant distance from the focusing coil. For mobile guns placed in the vacuum chamber, the Z movement is normally provided within the gun-traversing mechanism so that the main manipulating system merely rests on the solid base of the chamber, as can be seen in Fig. 4.2.

Optical viewing systems

Nearly all EBW machines are provided with optical viewing systems of varying degrees of sophistication. The few that do not incorporate such systems are often special-purpose mass-production machines which, once set to perform a given welding task, will require no further adjustment. The importance of a viewing system can be appreciated once it is recalled that the workpiece and its manipulating device are housed in a vacuum chamber, and the operator's eye is generally some distance away from the weld line looking through a port in the chamber. Again, because of the long working distance between the gun and the workpiece, the relative position of the point of impact of the beam and the weld line cannot be assessed with the naked eye; some optical aid is essential.

The optical viewing system performs a number of functions.

COMPONENT ALIGNMENT. Before welding, the component is aligned below the gun so that the beam would impinge on the weld line throughout the complete weld cycle. The operator thus sets up the component and energizes the manipulator drive, viewing the weld line through the optical device and adjusting the workpiece position as necessary. This is a critical operation because of the precision nature of welding by electron beam.

BEAM FOCUSING. This is carried out in most machines by viewing the size of the heat spot generated on a tungsten or copper block before welding. It is true that, by viewing through a port in the chamber, some degree of focusing can be achieved, but, since the size of the focused beam is a critical welding parameter, such focusing has to be performed with great accuracy. A form of optical aid is indispensable.

VIEWING DURING WELDING. Although welding speeds in EBW are too high for corrective action to be taken during welding in the case of a mishap, the viewing device is frequently used to observe the welding operation.

Following the original Zeiss system, high-voltage electron-beam columns have always incorporated an optical system coaxial with the electron beam itself, as in Fig. 4.3. A tungsten tube surrounds the beam shielding the optical components. The optical system is protected from metal-vapour contamination by a shutter and a glass shield. A field some $\frac{1}{2}$in. (13mm) across can be viewed at magnifications of up to $\times 40$. Zoom optics are used so that no adjustment is required following a change of magnification. Binocular arrangements are naturally preferred to monocular ones since they are easier for the operator. The optical graticule is provided with the cross-wires required for component alignment, the intersection representing the point of impact of the beam. A calibrated scale or grid is also provided in the viewing system which allows the operator to measure distances accurately at the workpiece surface.

Polarized filters are sometimes provided to observe the welding operation and to assist in reducing the effect of the brightness of the tungsten block during high-power beam focusing.

Two aspects of the illumination of the system deserve special attention. Firstly, the illumination must be powerful enough to meet the requirements of welding at long distances from the gun. Secondly, bulb replacement should be a rapid and simple operation, perhaps by arranging for the bulb to be housed on the outside of the electron-beam column.

When guns are located inside the vacuum chamber, the system described above is difficult to apply, mainly because of the mobility of the gun. An externally mounted viewing telescope is then used, aligned with a mirror mounted on the gun to reflect the area of impact of the beam. The mirror has to be protected by a shutter from the metal vapours and spatter generated during welding.

As will be described later in this chapter, closed-circuit television systems can be used in conjunction with the optical viewing system.

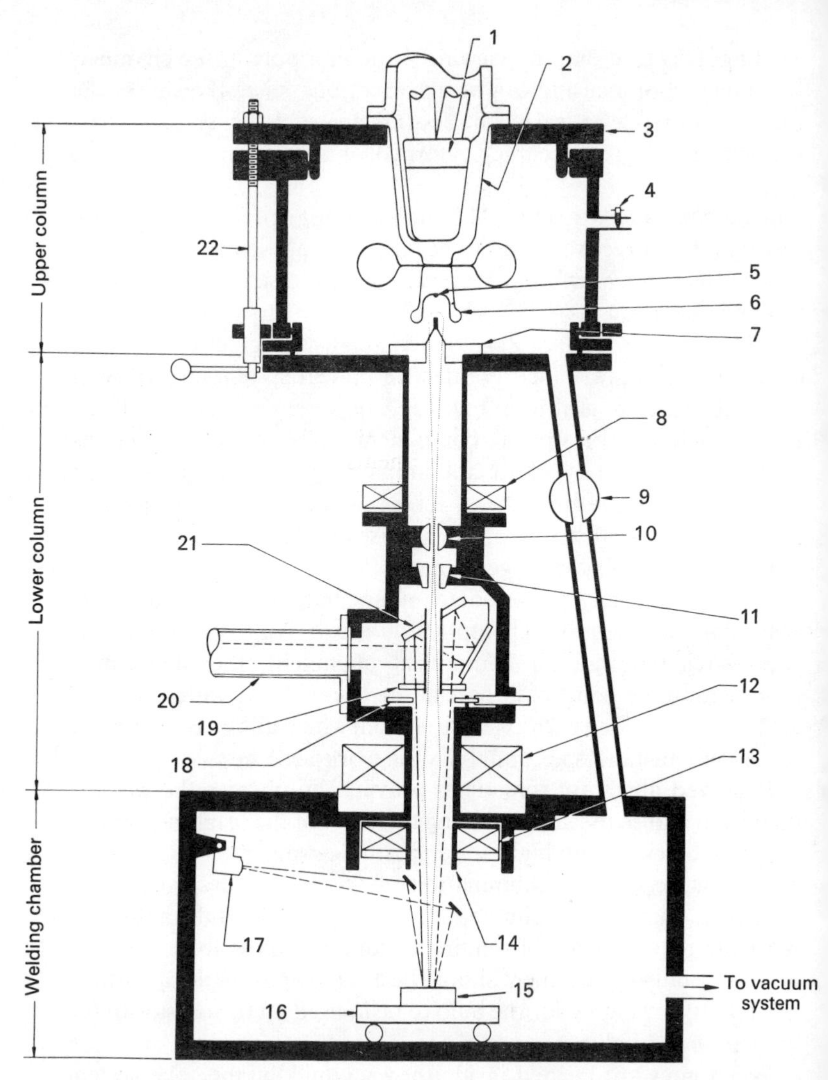

Fig. 4.3 Hamilton-Zeiss cutting/welding machine with optical viewing system: 1—demodulator and RF transformer; 2—cathode insulator; 3—cathode-adjusting plate; 4—manual air-inlet valve; 5—filament; 6—grid; 7—column anode; 8—stigmator coil; 9—column bypass valve; 10—column shut-off valve; 11—lower diaphragm; 12—magnetic lens assembly; 13—deflection coil; 14—heat shield; 15—workpiece; 16—worktable; 17—spotlight; 18—optical shutter; 19—vapour shield (glass); 20—optical viewing system; 21—block and mirror assembly; 22—adjusting shafts (*Courtesy Hamilton Standard Division of United Aircraft Corporation*)

In the absence of an adequate viewing system, some users resort to a simple technique: the gun is aligned as closely as possible with the weld line, a very low-power beam is switched on, and the component manipulated below the beam. Limited surface melting amounting to no more than a scribe mark is produced which is related to the weld line. A simple viewing device will then reveal any setting error and the necessary corrective action can be taken.

Soft-vacuum systems

In the early days of EBW, the chamber was evacuated to the same level of vacuum as that required to generate an electron beam, namely 10^{-4} torr. In fact, both the electron column and the chamber were evacuated by the same pumping system. The pump-down time was not excessively long, since the chambers were small in those days and the early users were generally experimenting with the technique and handled comparatively small numbers of parts. Thus, the evacuation time did not prove unduly embarrassing.

Two recent trends, however, have changed the situation dramatically. Firstly, applications have now extended to cover really large components demanding very large chambers, with the obvious penalty of either very high cost of large diffusion pumps or extended evacuation time, or both. Secondly, certain mass-production industries such as the motor-car industry could find applications for comparatively small pieces like gears. It is true that a small chamber would need only a short time for evacuation in such cases, but these industries require a mass-production type of machine where nearly all the operations in the welding cycle are automatic. Both situations have been met by the introduction of soft-vacuum EBW in which the gun is separately evacuated to the required beam-generating pressure of 10^{-4} torr while the chamber is held at a much higher pressure of, say, 10^{-1} to 5×10^{-2} torr.

An immediate consequence of increasing the pressure in the vacuum chamber is to increase the impurity level of the welding atmosphere. A hard vacuum is chemically very pure: at 10^{-4} torr, for example, the impurity level is 0·15 parts per million. Only a small number of applications, for instance the welding of reactive metals, require such a level of purity. Most welding operations utilizing arc methods operate under a shroud of an inert gas like argon. The level of impurity in commercially available argon is 500 parts per million.

It can be seen immediately that there is no real need for most applications to operate at a vacuum better than $3{\cdot}8 \times 10^{-1}$ torr, which corresponds to an impurity level of 500 parts per million.

There are, however, other facets to the concept of soft-vacuum welding. What, for example, is the effect on the beam itself now that the electrons are diffracted by the denser atmosphere, and what are the resultant welding characteristics with such a beam? The diffraction of the beam has two immediate effects. Firstly, focusing becomes more difficult, so that there is a drop in the maximum power density which can be achieved from a given electron-optical system; this will naturally affect the penetration capabilities of the beam. Secondly, the working distance between the gun and the workpiece is reduced, i.e., the focal length of an electron lens is reduced in a higher pressure environment. Of course, both effects are closely related to the level of vacuum and, as described in chapter 3, both features, namely deep penetration and long working distance, are not unduly affected by a soft-vacuum environment.

There is also the practical requirement of maintaining a stable differential pressure between the electron column and the chamber. This is achieved by providing the electron column with its own diffusion pump supported by a mechanical pump, while the chamber is held by a mechanical pump only. It is essential that the connection between the two units is of a low-conductance vacuum path, such as a small orifice, only sufficient to allow the beam to go through, but providing adequate resistance to the passage of low-pressure air from the chamber to the electron column. Naturally, the more powerful the diffusion pump connected to the column, the easier it is to maintain the dynamic pressure equilibrium of the system. Other factors also influence the situation, such as the welding pressure in the chamber which depends on its size and the capacity of the chamber's own pumping system.

We have seen that the welding advantages of a vacuum electron beam can be substantially maintained under soft vacuum conditions, and that it is possible to construct a vacuum system capable of dynamic equilibrium. Let us now examine the practical realization of soft-vacuum equipment.

Mass-production machines

'Single shot' machines are now commercially available and are designed mainly for welding small components. One component is

Fig. 4.4 Single-shot machine with twin manipulators (*Courtesy Sciaky Electric Welding Machines Ltd*)

Fig. 4.5 Single-shot machine used for gear welding (*Courtesy Hawker Siddeley Dynamics, Electron Beam Division*)

handled at a time and the vacuum chamber houses it, its fixture, and manipulator. These displace most of the volume of the chamber so that the remaining space is evacuated to the soft-vacuum level in a few seconds or even a fraction of a second. Basically, a two-valve system is employed which is cycled automatically. The object is to maintain the hard vacuum in the electron column at all times while the work chamber is vented to atmosphere for loading and unloading. While welding, the valve separating the column from the chamber is open and the loading valve in the chamber is closed. The latter is of a self-locking type designed to withstand the differential pressure loads on it. At the end of the welding sequence, the separating valve is closed and the chamber vented to atmosphere. Machines are designed so that all operations are sequentially controlled and interlocked for rapid and automatic cycling. Thus, a single action such as feeding the chamber to the column, operates the valves and

Fig. 4.6 Two-gun, single-shot unit (*Courtesy Hamilton Standard Division of United Aircraft Corporation*)

the pumping system, actuates the manipulator, and switches the beam on and off with power slope-in and fade-out as required. In fact, the general layout and operational sequence are not unlike those in high production resistance-welding machines.

In one design, shown in Fig. 4.4, two identical manipulators are used. While one component is being assembled on the first, the welding operation is proceeding with the other. When this is complete, the mechanism holding both manipulators is lowered, swinging through 180°. The welded part is removed, the loaded manipulator is fed into the column, and so on. It is obviously feasible for this machine to be equipped with other automatic feed systems such as a dial feed table. Another machine is shown in Fig. 4.5 where no automatic feed device is used; the tooling module is shown detached from the machine. Two welding columns can of course be used, as in Fig. 4.6. While one column is in the pump-down/weld/vent sequence, the tooling associated with the second column is being loaded or unloaded. This arrangement is particularly suited to applications in which the welding cycle is long, as the component flow rate can be nearly doubled.

Such mass-production machines are normally acquired to meet a specific production requirement. Naturally, the power of the electron gun and its other basic characteristics are selected to deal with a specific welding operation. Guns varying in power rating from 3 to 30kW are commercially available, with a range of accelerating voltages.

Naturally there are differences in the design concepts of the various manufacturers. However, in soft-vacuum systems, the separating orifice is situated at the first cross-over point of the beam after generation, and the focusing coils are placed in the soft-vacuum zone.

Machines with large chambers

The principle of soft vacuum can also find application in non-mass-production areas, particularly when the chamber is large. It is still possible to maintain the required dynamic pressure equilibrium by introducing a low-conductance vacuum path between the gun column and the chamber. The trend at present is towards incorporating such an arrangement as a standard unit, even though a diffusion pump may be used in evacuating the chamber to a hard vacuum when required. When a soft vacuum is adequate, the diffusion pump is isolated and the chamber is evacuated quickly by means of the mechanical and booster pumps only. An example of a recently constructed unit is shown in Fig. 4.7.

Fig. 4.7 Large EB machine with soft-vacuum capability (*Courtesy Hawker Siddeley Dynamics, Electron Beam Division*)

Out-of-vacuum systems

Large chambers of immense dimensions can be built and the soft-vacuum concept reduces both capital cost and evacuation time. However, there will always be a class of application where the vacuum chamber concept is almost completely irrelevant: for example, continuous welding of pipes in a steel mill, the welding of ships, or the construction of large atomic reactor shells. An out-of-vacuum system becomes essential provided EBW does not lose its basic features by bringing the beam into an atmospheric pressure environment.

It is well appreciated that the lower the accelerating voltage, the lower the speed of the electrons and the more rapidly a well-focused beam becomes diffused when it passes through a gas at atmospheric pressure. Thus, as the voltage is increased, it is more likely that a significant power density is attained at greater working distances. At least 120kV are required for a practical working distance and better results are obtained if the voltage is increased up to the practical level of 200kV. It must be appreciated that there is also a power loss when the beam is diffracted by the dense environment of

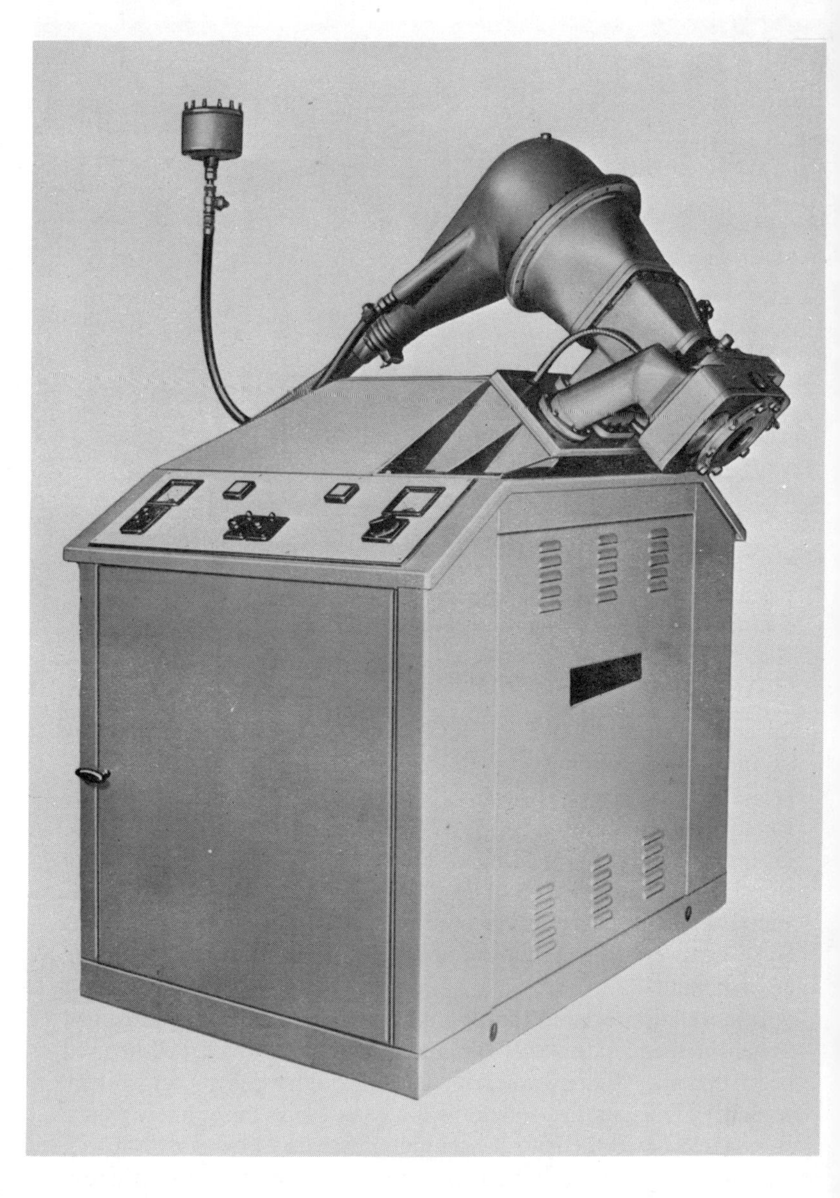

Fig. 4.8 Early Heraeus non-vacuum EBW machine (*Courtesy Leybold-Heraeus GmbH*)

Fig. 4.9 Deep penetration by non-vacuum electron beam in stainless steel (*Courtesy Hamilton Standard Division of United Aircraft Corporation*)

a gas at atmospheric pressures. Guns with powers of $7\frac{1}{2}$, 12, and 25kW have been successfully used 'out of vacuum'.

Reference has already been made to the early work of Heraeus in building out-of-vacuum electron-beam guns; see chapter 1. Figure 4.8 shows an illustration of one of the few machines built as early as 1955 (acceleration voltage 200kV; power 2kW). More intensive effort has been carried out recently by Hamilton Standard aimed at determining the capabilities and limitations of a non-vacuum electron beam. The results are described in detail in chapter 3. It is sufficient to state here that both the working distance and the nature of the shrouding gas have a significant effect on the penetration capabilities of the beam. However, accepting the limitation of a short working distance, the deep penetration phenomenon of a vacuum electron beam is still present in a non-vacuum situation, although its magnitude is significantly reduced. This is illustrated in Fig. 4.9 where a butt weld in 0·28in. (7mm) stainless steel has been carried out. Although the depth-to-width ratio of the fusion zone is not as great as that normally experienced when welding in a vacuum, the weld configuration reveals that the deep penetration mechanism has been operative.

Thus, high welding speeds can be achieved in an out-of-vacuum operation as compared with arc-welding methods; the heat input is

Fig. 4.10 Hamilton Standard non-vacuum gun for tube welding (*Courtesy Hamilton Standard Division of United Aircraft Corporation*)

also lower. The limitations of the short working distance restrict non-vacuum EBW to shapes of simple contours, and welding in 'inaccessible' areas cannot be achieved. It must also be remembered that, besides gas shielding to avoid metal oxidation during welding, X-ray shielding is also required; there is now no vacuum chamber to absorb the X-rays generated when the beam hits the target workpiece.

The non-vacuum machine shown in Fig. 4.10 was used to weld stainless steel tubes in a continuous tube mill. Figure 4.11 illustrates the arrangement for welding a spherical steel shell.

In outer space, there is no air and EBW obviously has great potential; a prototype 'hand held' gun designed for use by astronauts can be seen in Fig. 4.12. This gun is rated at 1kW (80kV, 12·5mA) and has the capability to weld steel up to 0·125in. (3mm) thick. It is conceivable that further developments will permit greater thicknesses to be welded and ultimately space stations may well be fabricated by EBW in outer space.

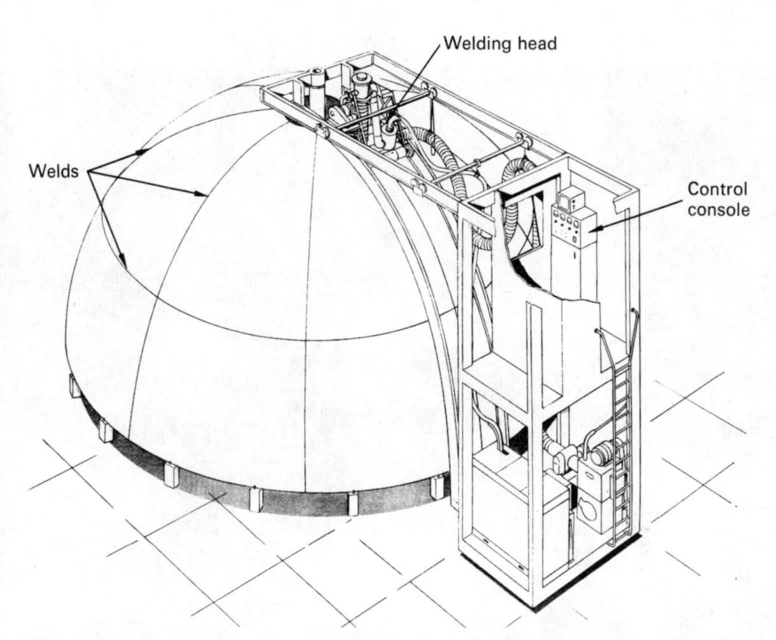

Fig. 4.11 Concept for shell welding by non-vacuum beam (*Courtesy Hamilton Standard Division of United Aircraft Corporation*)

Machine accessories

Magnetic beam deflection

Two sets of magnetic deflection coils are normally supplied with most guns and are located below the focusing lens. The coils are used to deflect the beam with respect to the electron-optical centre-line in two mutually perpendicular directions. Depending on whether only one coil is energized or both simultaneously, the pattern of beam deflection will be either in one direction or in a co-ordinated motion. Again, the coils can be selectively supplied with either dc or ac signals. It is also possible to superimpose ac onto a dc signal. Let us now examine the circumstances where such facilities may be usefully utilized.

The most frequently used mode of beam deflection is a small-amplitude oscillation normal to the weld line at mains frequency. Thus, while the component is moving below the beam, say in a straight line, the beam is made to oscillate transversely. The amplitude of oscillation is normally small, say 0·010–0·015in. (0·25–

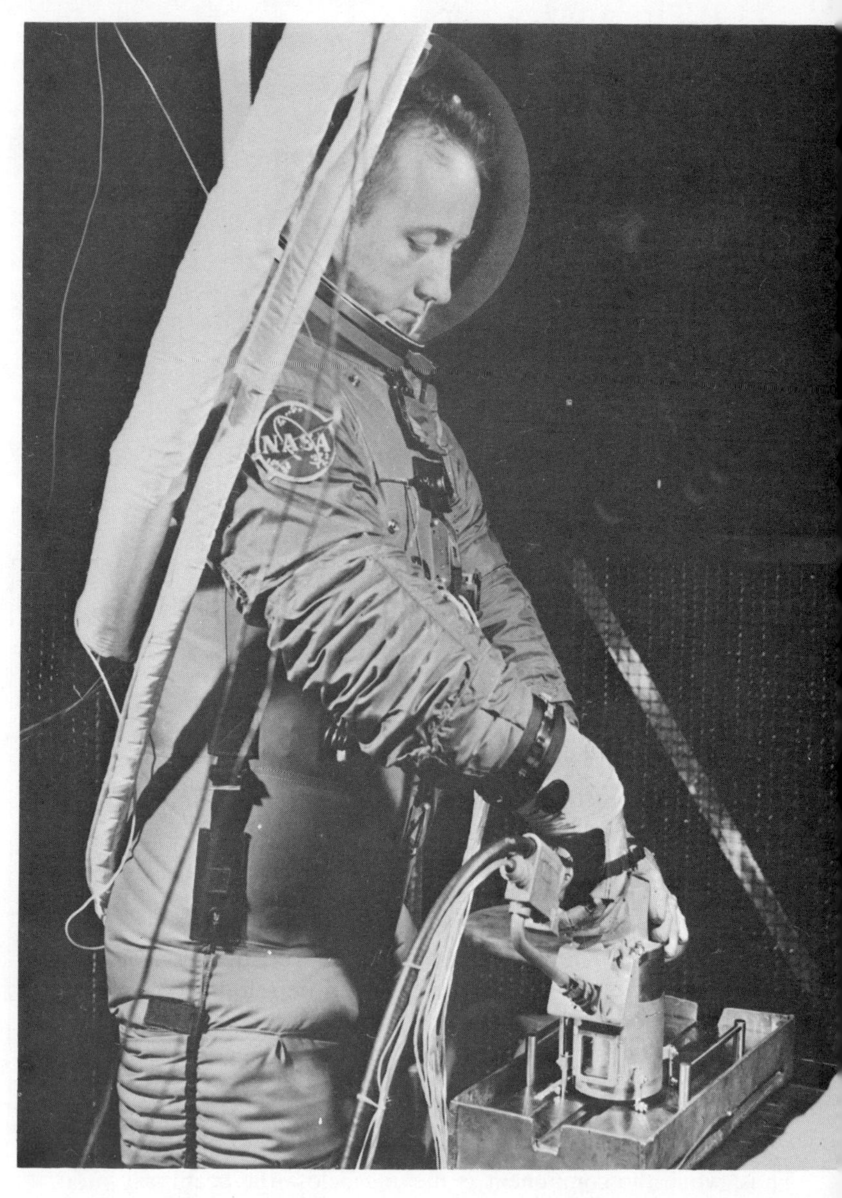

Fig. 4.12 Space EBW gun (*Courtesy Hamilton Standard Division of United Aircraft Corporation*)

0·40mm), requiring a beam deflection of no more than 5° from the gun centre-line, depending on the distance between the deflection coil and the surface of the workpiece. Such a facility is normally used to produce a weld wider than would be achieved from a static beam. It is particularly useful when welding thin-gauge material or when the use of a wider weld may relax the accuracy of machining of the abutment, provided the additional heat input to the component can be tolerated. When welding dissimilar materials, it is often observed that beam oscillation assists in the mixing of the two materials and improves the appearance of the upper seam of the weld.

The wave pattern is naturally controlled by the characteristics of the input signal to the coils. This can be a sine wave or a 'chopped' form as illustrated in Fig. 4.13; the pattern of deflection affects the heat distribution over the width of the weld zone.

When both sets of coils are energized simultaneously with a sine wave with a 90° phase shift, the beam is made to rotate at the

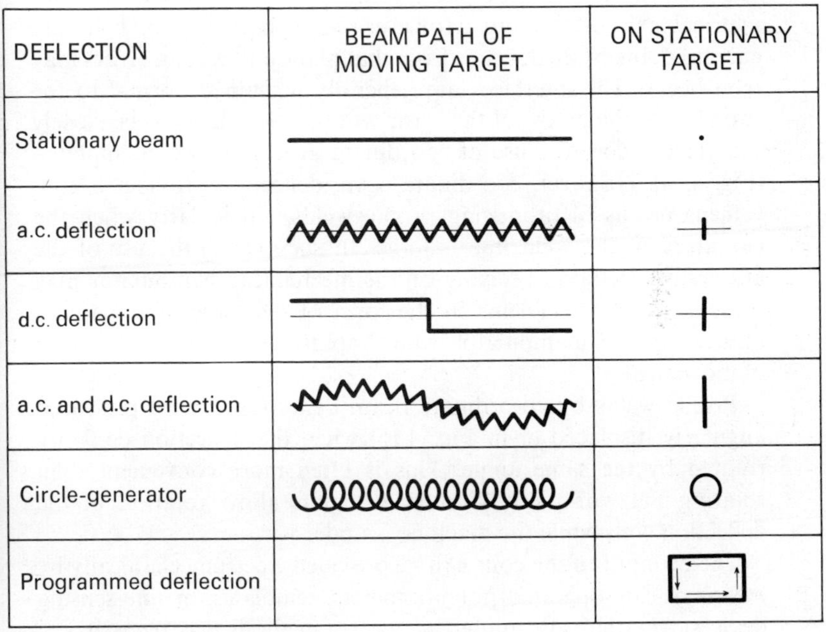

DEFLECTION	BEAM PATH OF MOVING TARGET	ON STATIONARY TARGET
Stationary beam		
a.c. deflection		
d.c. deflection		
a.c. and d.c. deflection		
Circle-generator		
Programmed deflection		

Fig. 4.13 Beam deflection patterns (*Courtesy Hamilton Standard Division of United Aircraft Corporation*)

frequency of the input signals about its main position. The pattern will be either circular or elliptical depending on whether the input to the two coils is of equal or differing magnitude. It has been reported that if the beam is made to oscillate at high frequency, say 10kHz, the incidence of porosity in the weld is substantially reduced, if not eliminated altogether, probably because of the stirring action in the molten pool.

The dc input to the coils provides a steady deflection of the beam at welding speeds. In such cases, welding can be performed without traversing the component. In effect, the gun becomes mobile but operates within the geometrical limitations of the deflection system. Even with specially designed wide-angle deflection coils, up to 30° in certain cases, welding by beam deflection is limited to an area 2in. (50mm) or so across. Nevertheless, the added degree of flexibility can be particularly useful in certain applications when combined with the movement provided by the mechanical manipulator.

When the beam is deflected by one coil in the direction of the weld seam, it will move in the plane containing the seam and the geometrical axis of the gun. However, when both sets of coils are energized, the beam deviates from that plane and certain errors may then occur. The interfaces are generally machined normal to the surface and the extent of the error will therefore have to be closely examined before the use of coordinate deflection is contemplated (Fig. 4.14). However, coordinate beam deflection provides a convenient means of producing profile welds, particularly when the curvature of the weld line is large. In such cases, the use of the alternative method of relying on the mechanical manipulator may present its own problems in the form of the large inertia forces induced in the manipulator while rapidly changing its direction of movement.

Profile welds by co-ordinate beam deflection can be produced angularly displaced (θ in Fig. 4.15) when the deflection coils are rotated by the same angle. This is often more convenient than rotating the workpiece. Certain EB guns allow rotation of the deflection coils while the machine is under vacuum.

The dc input to the coils can be provided electromechanically by means of cam-operated potentiometers, templates, or line-sensing devices. Alternatively, numerical control methods may be used.

It is possible to superimpose ac deflection on the dc co-ordinate signals, thus producing a wider seam in a profile weld.

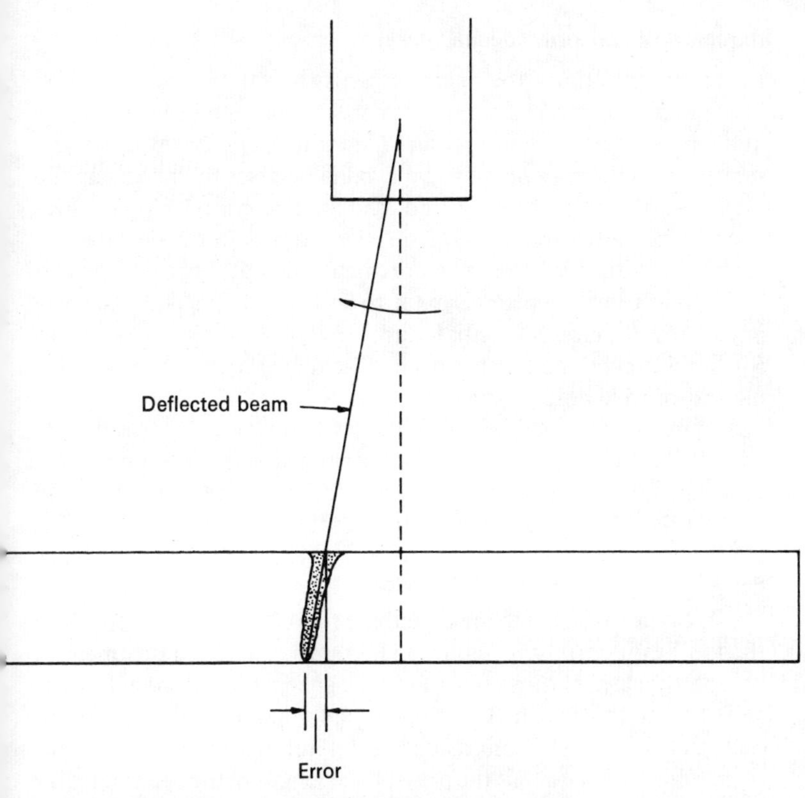

Fig. 4.14 Error resulting from excessive beam deflection

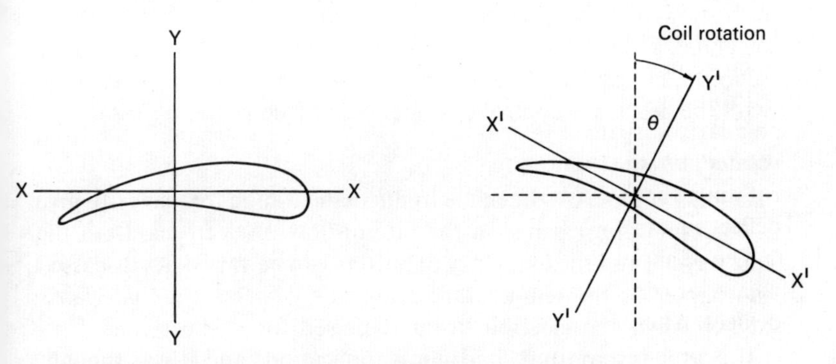

Fig. 4.15 Angular displacement of weld pattern by coil rotation

Automatic beam switching and decay

In most applications, the operator can switch the beam on and off as required. However, if the speed of welding is high or when a large number of operations are required, an automatic beam timer and sequencing unit may prove to be a valuable asset by increasing the accuracy of reproducibility and reducing operator fatigue. Frequently, it is advisable to 'slope-in' the beam power rather than to switch on suddenly; similar requirements may also exist at the end of the weld when the beam is made to 'fade-out'. This is particularly important in circumferential and orbital welding in order to control the length and penetration of weld overlap and to overcome the end-of-weld crater formation.

Timers are provided by which the beam power is increased at an adjustable 'slope-in' rate to a predetermined level at the start of a weld run, and decreased to nil after a set time at the end of the weld by an adjustable 'fade-out' rate. In circular welds, the beam may be controlled by revolution counters by cams attached to the rotary fixture.

It is often advisable to tack-weld the two component parts before the full beam current is applied. The timer provides adjustment of on and off periods thus controlling the spacing between tack welds and the length of each. The slope-in/fade-out control operates on each tack weld. In certain machines, an automatic focusing circuit is included to ensure that the preset focal length of the beam remains unaltered during changes in beam power.

In multispindle fixtures used for quantity production, the beam-power timer is linked to the drive motor so that a complete set of operations can be executed by operating the main switch of the machine. Thus, by a single push-button, the workpieces are set in rotation, indexed, and the beam is made to weld over a preset time with the necessary control of slope-in and fade-out as necessary.

Beam pulsing

The bias voltage can be made to fluctuate at high frequency to give a corresponding change in the rate of flow of electrons. Both the 'duty cycle' and the frequency of pulsing can be varied. As discussed in chapter 3, the temperature attained by the target material is higher when the electron beam is pulsed for a short time. This parameter is important in drilling applications and it was thought that even deeper weld penetration might be achieved with a pulsed beam because of the increased drilling action. However, although the power unit of some EB machines incorporates voltage control for pulsing, this facility is seldom used, as the effect of beam pulsing on weld penetration has not been universally demonstrated.

Angular gun column

Welding by an electron beam is accompanied by the generation of metal vapour in the centre of the weld pool. In the majority of cases, the quantity of vapour produced is so small as to have no detrimental effects. However, when welding thick sections of metals which produce a lot of vapour, such as aluminium and titanium, some of the vapour atoms find their way through the electron column and are deposited on the cathode filament. The emission characteristics of the cathode are thus substantially altered and it becomes extremely difficult to control beam power. In extreme cases, cathode/cathode-shield flashover takes place leading to the generation of excessive beam currents and possible damage to the workpiece.

Fig. 4.16 Angular gun column (*Courtesy Hamilton Standard Division of United Aircraft Corporation*)

Fig. 4.17 (a) Steigerwald 'dry' gun, and (b) diagram: 1—seal; 2—quick-change filament cartridge; 3—beam source; 4—magnetic lens; 5—deflection system; 6—column valve; 7—magnifying optical system; 8—vapour shields; 9—workpiece; 10—optical illumination; 11—diffusion pump (*Courtesy K. H. Steigerwald*)

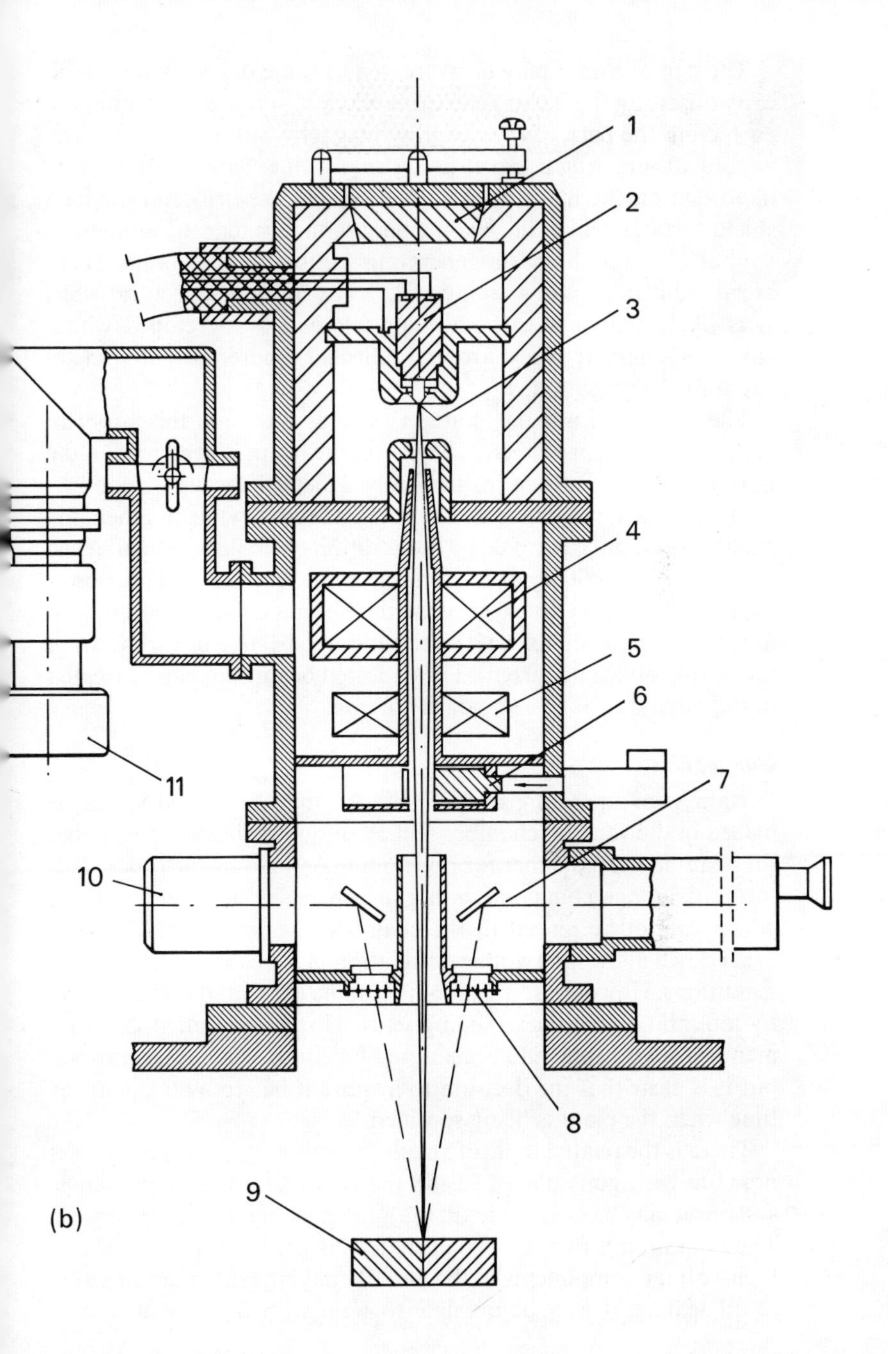
1
2
3
4
5
6
7
8
9
10
11
(b)

The complication may be overcome to some degree by mechanically offsetting the beam-generating section of the gun column and correcting the path of the beam by magnetic beam deflection. The vapour atoms, which travel in a straight line, would not then be deposited on the filament. Also, the smaller the orifice in the heat shield through which the beam emerges, the smaller the amount of vapour reaching the beam-generating section of the column. There are also advantages in working as far away from the gun as possible. It is likely that new cathode materials will be developed whose emission characteristics are not affected appreciably by metal vapour contamination.

The use of an angular column should overcome this problem completely. As can be seen in Fig. 4.16, the beam is generated in the normal way but is then magnetically deflected through some 20°. Certain guns are manufactured in such a layout, while in others the standard column is adapted by the addition of auxiliary components.

Besides the auxiliary deflection coil, some regulation of the power supply is necessary to ensure that the deflected beam continues to flow along the main axis of the column irrespective of variations in the beam voltage or current. The deflected beam can then be treated in the same way as in a standard column.

Gun mobility

Certain guns operating in the voltage range up to 60kV can be placed in the vacuum chamber and are mounted in such a way that they can be moved either for positioning or welding purposes. This is an advantage compared with guns operating at higher voltages which cannot be placed in the chamber because of the electrical breakdown in the high-voltage insulation of the cable under vacuum conditions. However, some mobility can be given to the gun column by mounting it on linear sliding seals. This arrangement is convenient for positioning and is also used for linear welding operations, but it is clear that the decision to acquire it has to be taken at the time when the plant is being specified.

There is the related topic of angular positioning of the gun. This is possible with guns placed inside the chamber. With high-voltage systems it may be possible to tilt the gun up to, say, 15° to the vertical. The limitation is due to the fact that oil is used in the insulation of high-voltage components in the gun. By paying particular attention to oil sealing, it may be possible to weld with the gun tilted at a greater angle. However, in a recent gun design, no oil is used for insulation (Fig. 4.17). Such a 'dry' gun can naturally be used at any angle, thus simplifying the design of certain fixtures and reducing their cost, for example, when welding two conical surfaces.

Wire feed

Satisfactory EB welds can be produced with a plain interface, but the control of welding parameters such as beam power and welding speed is critical if an acceptable geometry of the upper and lower beads is to be achieved. This is so since an underpowered weld would result in incomplete penetration, while one that is overpowered could produce channelling of the upper bead and 'drop-through' at the lower bead. The situation is less critical when additional metal is provided, for example, in the form of wire automatically fed into the weld pool.

Automatic wire feed can also be used to fill moderate gaps in the interface, thus lowering the cost of machining by reducing its tolerances. It may also open up the area of sheet metal fabrication to EBW which has hitherto been excluded since it is difficult to produce the required contact accuracy in sheet metal parts. In salvage applications, it is often necessary to add material to fill in gaps or build up worn parts. If the wire material is selected to increase the alloy's hot ductility, it may also have some desirable metallurgical effects as a means of reducing the tendency to cracking in the more complex alloys.

Wire-feed devices used in conventional welding are unsuited to EBW for a number of reasons. Generally, finer wires are required for EBW and the whole unit has to be precision-built to meet the accuracy of EBW operations. The following are the essential characteristics required for a successful EBW automatic wire-feed device:

The straightening mechanism must be capable of presenting the wire to $\pm 0{\cdot}005$in. (0·1mm) from the axis of the nozzle. If the wire deviates from the path of the beam it will not find its way into the molten pool.

The nozzle must be capable of remote control inside the vacuum chamber.

Reliable drive mechanisms are essential, since any slip in the drive would result in local metal starvation in the weld with variable weld quality.

Certain manufacturers supply automatic wire-feed devices which are normally fixed to the heat shield of the gun column. The various parameters such as nozzle alignment, height, and speed of wire feed are controlled from a unit outside the chamber. The drive motor is placed inside the chamber but, to avoid the resulting encumbrance in the region of the workpiece, the motor may be placed elsewhere and the power transmitted through a flexible transmission shaft. Figure 4.18 shows a commercially available wire-feed device.

Fig. 4.18 Automatic wire-feed device (*Courtesy Sciaky Electric Welding Machines Ltd*)

Closed-circuit television viewing

The object of using a closed-circuit TV system is to permit visual monitoring of the weld joint line for setting-up purposes, thus reducing operator eye fatigue which results from working for long periods with the optical system. The TV system is particularly useful when the machine is used for repetitive production tasks involving manual control of the workpiece-indexing fixture. The setting-up operation can be viewed simultaneously by a number of people, unlike the optical system which is normally available only to the operator. The task of supervision and inspection are therefore facilitated.

Television viewing is rapidly becoming popular with both fixed external guns and mobile internal guns. In fixed guns with a telescope viewing arrangement, the camera replaces the eye of the operator. A filter is normally inserted between the camera and the telescope to protect the Vidicon tube from excessive brightness, e.g., when a high-power beam is being focused on a tungsten block. When guns are placed inside the chamber, the camera is attached to the electron gun and moves with it. By means of a shutter-protected mirror and lens system, the camera views the workpiece coaxially with the electron beam. The Vidicon tube is specially designed to operate under vacuum conditions. A recent example of such a system can be seen in Fig. 4.19. With highly reflective workpiece surfaces, it may prove useful to finish-machine, or polish, each side of the joint in different directions; the weld line would then appear as the junction between two zones of differing brightness.

Fig. 4.19 Closed-circuit television screen (*Courtesy Sciaky Electric Welding Machines Ltd*)

Seam-tracking techniques

One of the main obstacles in the way of a wider employment of EBW in industry is the necessity for very accurate alignment between the beam and the joint to be welded. Initial alignment may be achieved by the use of one of the several alternative optical viewing systems available, but it is in the nature of sheet metal parts that perfect alignment cannot always be secured. Even in forged and machined parts, the joint may not remain aligned during welding owing to progressive component distortion. Although this is very small with EBW, some movement nevertheless does take place and may be sufficient to cause misalignment errors resulting in an unsatisfactory weld. This is particularly so when a large number of adjacent welds are to be made and the distortion becomes cumulative.

Whatever the reason for the discrepancy between a nominal weld path and the actual weld line, correction must be rapid and precise. It may appear attractive to make such a correction by beam deflection, mainly because the beam is virtually of no inertia and the correction can be applied very quickly. However, if the change of beam angle to accommodate the error is large, or if the section to be welded is thick, or both, the lower part of the interface may be missed (see Fig. 4.14), although the beam may still be accurately focused on the top of the joint line. It is preferable, therefore, to make the correction while the beam remains normal to the surface of the workpiece. This is done by utilizing mechanical movement of either the worktable or, if it happens to be mobile, the gun.

The weld line can be maintained in the path of the beam by employing a self-adaptive seam-tracking control system. The seam is followed exactly by a tracking device which registers any deviation from the nominal path. The error signal is fed to a servomotor which applies the necessary correction.

Let us consider the case of a linear weld produced by manipulating the worktable in a straight line. In the absence of a seam-tracking system, the optical device is used to align the starting point of the weld line below the beam. The drive is then energized and the complete length of the weld line is viewed to see whether it is maintained below the gun axis. If it is not, the component is realigned and the procedure repeated until true alignment is achieved. Even then, this presupposes that the workpiece has been prepared very accurately.

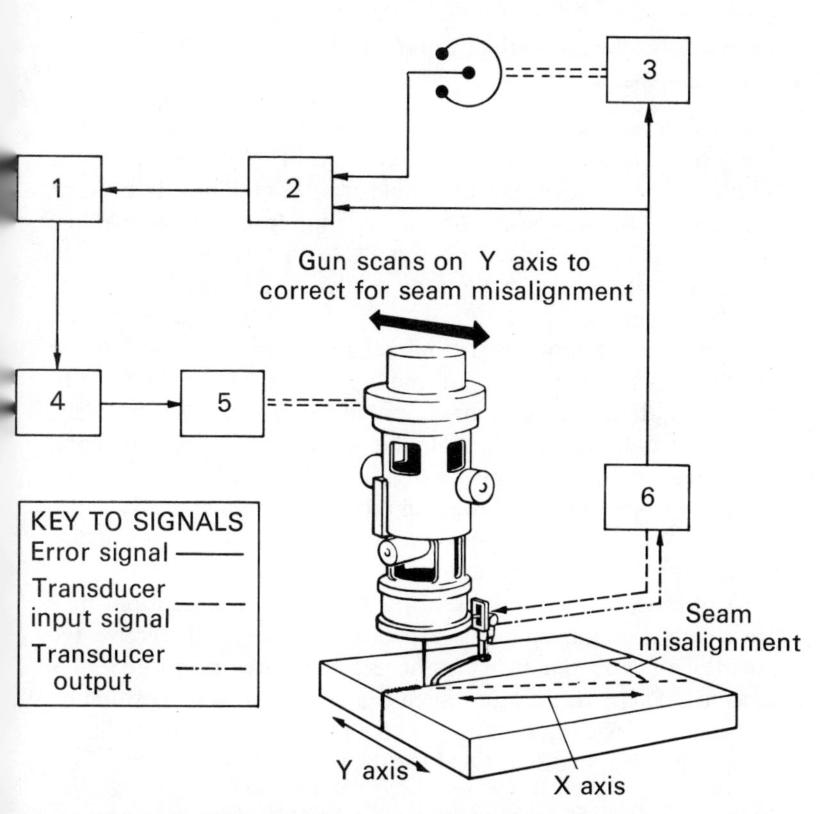

Fig. 4.20 Diagrammatic illustration of seam-tracking system: 1—phase shift circuit; 2—dc amplifier; 3—servo-amplifier; 4—armature power shift; 5—dc motor for Y axis drive; 6—oscillator (*Courtesy Sciaky Electric Welding Machines Ltd*)

When a seam-tracking system is employed, it is not essential for the contact to be an exact straight line, nor is it necessary for the alignment to be completely true. The tracking system would sense any errors and apply the necessary correction. It is clear that this reduces both the cost of machining and the time consumed in setting up the component in the machine.

The basic components of a seam-tracking system are shown in Fig. 4.20 and include:

the seam-tracking transducer;
the oscillator energizing the transducer;

a servomotor for both velocity and position;
a control panel.

It will be seen from the illustration that the workpiece is driven along the X axis and the gun traverses in the Y direction. As the misaligned seam moves along the weld axis, the oscillator/transducer network determines the amount of deviation from the nominal line and feeds back the correction signals to the servosystem, which will reposition the electron gun by displacing it in the Y direction. It will be appreciated that this is a closed-loop control system and that the various motions are interchangeable. It is possible, for example, to weld with the gun moving in the Y direction and to correct by adjusting the X drive on the manipulator. Alternatively, in the case of fixed guns, the weld line will be in the X direction and the correction applied by displacing the table in the Y direction. The same principles obviously apply in the case of circular welds.

Alternative tracking devices can be employed which use a stylus, proximity transducers, or cam-following devices.

Seam-tracking methods are not widely used at present but they are a valuable aid, and it is expected that they will receive the attention of users in industry as the technique becomes more widely accepted. There are certain difficulties associated with positioning the tracking device close enough to the point of impact of the beam. Also, tracking systems are costly at the present time.

5. Metallurgical considerations

This chapter deals with those aspects of EBW that are related to the metallurgical characteristics and mechanical properties of the material. The essential question raised when considering a welded structure is whether the fabrication will give adequate service performance. This depends on both the service conditions and the joint properties as determined by the presence of defects and weld shape. R. G. Baker[1] presented the same argument by stating that 'A steel showing ideal weldability should be capable of being joined by any desired welding process, without special precautions, to produce joints whose properties allow the full potential of the steel to be utilized'. Ideally, therefore, welding should result in no deterioration of joint properties including tensile strength, corrosion resistance, creep strength, and ductility. It is unlikely that any engineering material when welded by any process would fulfil these ideal requirements. In practice, fusion welding does change the metal properties by the act of melting and solidification, together with the associated thermal effects resulting in metallurgical changes in the parent material, and some geometrical changes due to the weld shape.

Electron-beam welding can justly be said to approach the ideal fusion welding process as it results in the minimum thermal disturbance in the workpiece for a given depth of weld. However, even with EBW, as it is still a fusion welding process, changes in the weld area still occur and the resulting weld has to be assessed in terms of the effect of the imperfections introduced by welding on service requirements. Electron-beam welds are subject to the same basic difficulties as any other fusion welding process including cracking, porosity, lack of fusion, control of bead shape, hardening, and softening. This chapter will be concerned with the occurrence of these defects together with techniques for their control. It also discusses the effect of these defects on joint properties. However, before dealing with specific materials, it is worth reviewing the metallurgically advantageous features of EBW which are made use of in practical welding, together with the disadvantages.

The transfer of heat from a line source (as opposed to a point source with conventional fusion welding processes) through the full depth of fusion results in the minimum heat input to the workpiece. This situation has two consequences of major importance. Firstly, distortion is held to a minimum, since the combined elastic and plastic deformation in the workpiece is related to the total heat input. A further effect of the line heat source and parallel-sided fusion zone is that any thermal expansion and contraction occurs almost uniformly through the full depth of penetration. Secondly, the low heat input to the parent material confines the metallurgical changes to a narrow band on either side of the fusion zone. The width of this band varies from one to two times the fusion zone width depending on the material type and thickness, and on the welding conditions. As EBW is a fusion welding process, the same metallurgical effects occur in both the parent material and fusion zone as in any other fusion process, although the associated heating and cooling rates and temperature gradients are very often much higher in the case of EBW.

The thermal strain and metallurgical effects are always interrelated, since it is impossible to heat a metal locally without producing some strain both during heating and cooling; the extent of the strain introduced depends on the heat input to the workpiece. Similarly, whenever a metal is heated, metallurgical changes occur, again during both heating and cooling, and the presence of strain can have an effect on the metallurgical changes which are occurring primarily as a result of thermal effects. Thus, thermal strains and metallurgical effects have to be considered together.

The reduced thermal strains and metallurgical effects have enabled a great many 'unweldable' materials to be satisfactorily joined by EBW including several dissimilar metal combinations. In addition, the low distortion aspect has allowed EBW to be used to weld fully machined components within acceptable distortion limits. In the following sections these effects will be examined in greater detail.

Advantages of electron-beam welding

Weld purity

One of the advantages resulting from the use of a vacuum environment in EBW is the absence in the weld atmosphere of oxygen, nitrogen, and hydrogen, which are normally present in arc-welding processes. Reactive materials are therefore particularly suited to welding by an electron beam. Figure 5.1a shows an electron-beam weld in a tantalum alloy where no oxidation and no gas absorption took place during welding; the joint therefore exhibited high ductility. A similar joint produced by vacuum-purged TIG-welding shown in Fig. 5.1b exhibited poor ductility owing to the large volume of recrystallized metal and unacceptably high gas absorption. It is interesting to note that the potential contamination level in the best grade of commercially available inert gas is some 1000 times as high as that present at a vacuum of 10^{-4} torr, the normal operating vacuum for EBW.

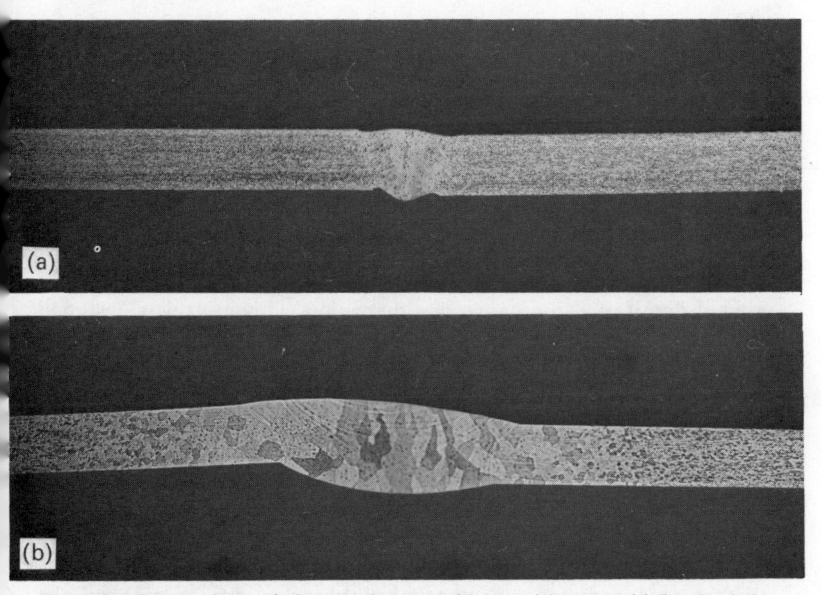

Fig. 5.1 Comparison of electron-beam weld (a), with arc weld (b): both in tantalum alloy (× 7) (*Courtesy Rolls-Royce Ltd*)

Low distortion

A comparison of the fusion and HAZs (heat-affected zones) in a typical EB weld with a TIG-weld in $\frac{1}{2}$in. (12·7mm) thick stainless steel plate is illustrated in Fig. 5.2. The volume of metal affected is

Fig. 5.2 Comparison of TIG-weld (top) with EB weld (bottom) both in $\frac{1}{2}$in. (12·7mm) stainless steel plate (*Courtesy The Welding Institute*)

approximately $\frac{1}{10}$ and it may be inferred that the thermal strains are of a similar ratio. The resultant low distortion and low residual strain field is probably the main advantage of EBW over other fusion processes. In the early days of the use of EBW, the aero-engine industry found this feature extremely valuable. Figure 5.3a illustrates a weld detail of a main aero-engine casing. This joint was designed to be produced by TIG-welding, but excessive distortion took place and EBW was used instead. Distortion after EBW was only one-seventh of that experienced on TIG-welding. A section through this weld is shown in Fig. 5.3b. Another example of low distortion is illustrated by the gear assembly shown in Fig. 5.4. The gears, in a Ni–Cr–Mo steel, were welded after final grinding and were found to be dimensionally accurate to within 0·002in. (0·05mm) after EBW. An examination of typical electron-beam welds shows that, of the distortion introduced, most is attributable to solidification shrinkage normal to the weld interface. Since an electron-beam weld is

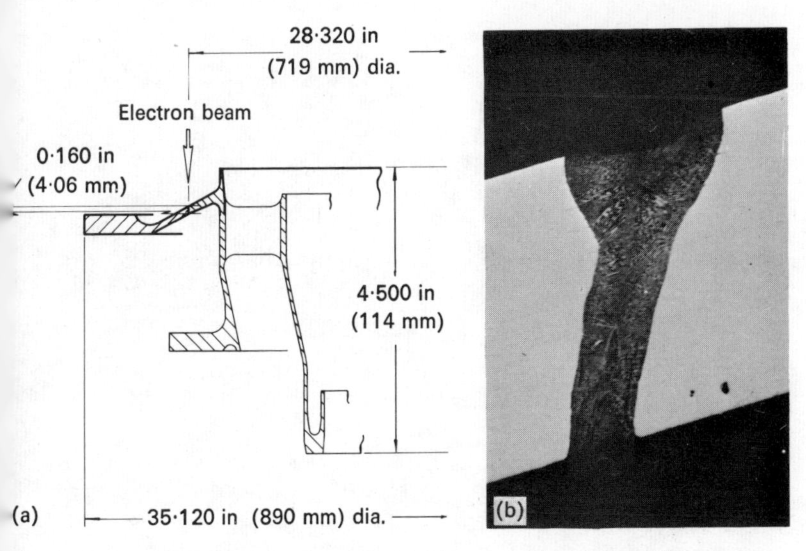

Fig. 5.3 Welding of flange to casing: (a) location of weld, and (b) macro-section of weld (× 10) (*Courtesy Rolls-Royce Ltd*)

essentially parallel in cross-section, weld shrinkage will be more or less uniform across the weld interface. In many instances, this shrinkage can be readily allowed for. It is advisable to carry out a quantitative assessment of the magnitude of the shrinkage displacement on a simulated component, as it is difficult to give shrinkage figures to cover the entire range of joints likely to be encountered. This is because joint geometry, the depth of penetration, and the material to be welded, all affect the degree of shrinkage. Generally, however, typical shrinkage displacements are between 0·005 and 0·010in. (0·1–0·25mm) for a weld width of about 0·050in. (1·2mm) and a weld penetration of between 0·100 and 0·500in. (2·5–12·7mm).

Metallurgical effects

The reduced thermal strain associated with EBW, together with the narrow fusion and HAZs, lead to two main metallurgical advantages: reduced incidence of cracking during and after welding, and improved weld properties because of reduced metallurgical damage.

The occurrence of cracking during welding either in the fusion zone or HAZ is due to two basic causes: the existence of strain

Fig. 5.4 Gear assembly; electron-beam weld shown in section of both gears (*Courtesy GEC Power Engineering Ltd*)

and the presence of a metallurgical structure which is unable to withstand this strain without rupture. This metallurgical condition can occur at any temperature and may be due to the presence of liquid films at high temperatures near to the melting point of the bulk metal or to a low-ductility solid phase at temperatures substantially below the melting point. The advantage that EBW offers in this context is that the strain field which needs to be relaxed is quite small.

Characteristics of the process

Within the deep-weld category of EB welds, it is possible to produce a range of weld widths and depth-to-width ratios which are controlled by the beam parameters, in particular, beam power, power density, and weld speed. A typical range of weld shapes is illustrated in Fig. 5.5 which shows transverse weld profiles in a 0·06in. (1·5mm) thick nickel alloy. The weld width and shape affect not only the strain field but also the cooling rate. This is an important factor since it affects both the metallurgical structure and the time during which the metal may pass through certain critical temperature ranges. The high cooling rate associated with EBW can have either good or bad consequences, which will be discussed later.

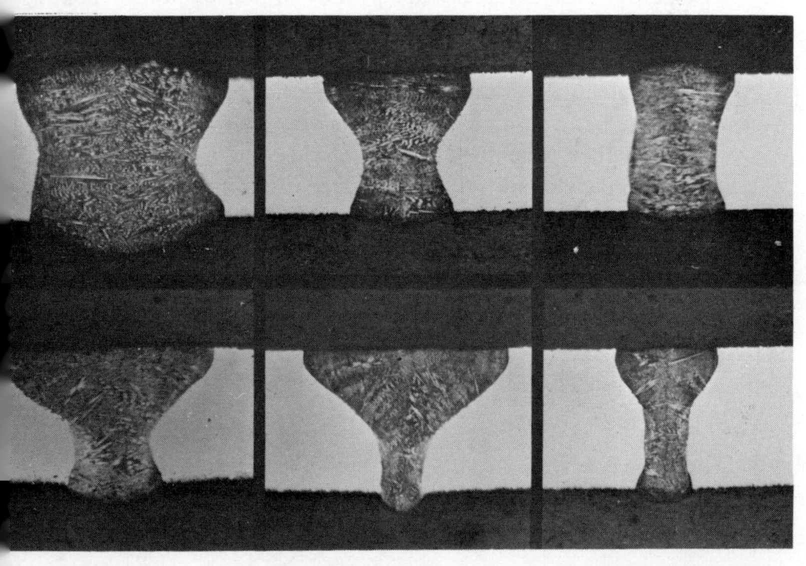

Fig. 5.5 Range of weld profiles in a 0·06in. (1·5mm) thick nickel-base alloy (×7) (*Courtesy The Welding Institute*)

The extent of heat spread into the workpiece controls the extent of the associated strain field. In general, the narrower the strain field the less chance there is of cracking. Narrow welds will produce high rates of induced strain which might be expected to increase cracking tendencies, whereas the narrow strain field reduces the risk of cracking. Whichever effect is more important probably depends on the material, particularly its strain rate sensitivity.

It is difficult to generalize as to which range of welding parameters is appropriate for a family of alloys that would produce defect-free welds. Even within one group of materials, it is sometimes found that high travel speeds are beneficial while in others low travel speeds are appropriate. It appears that, where it is possible for a small amount of grain-boundary liquation to occur in the HAZ, high travel speeds are beneficial because of the rapid thermal cycle which reduces the occurrence of liquation. Where substantial grain-boundary liquation is inevitable, however, it is probable that lower travel speeds are useful in producing more liquid, thus encouraging crack healing by liquid flow.

'Waspaloy' is a complex nickel alloy which exhibits cracking when TIG-welded, either immediately after welding or on post-weld heat treatment. The cracking (Fig. 5.6a) during welding was found to be associated with grain-boundary liquation, the extent of which, it was realized, increased with lower welding speeds. Electron-beam welding was then attempted at high speeds and was found to give a grain boundary condition almost free from liquation (Fig. 5.6b). When the EB parameters were altered to allow welding at speeds of less than 10in/min (4·2mm/s), substantial grain-boundary liquation was then observed. By welding at the higher welding speed

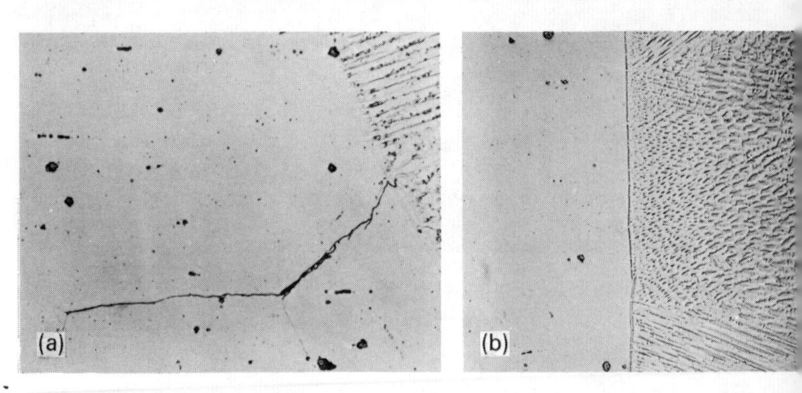

Fig. 5.6 Cracking in Waspaloy: (a) by grain boundary liquation, and (b) which is eliminated by high welding speed (×250) (*Courtesy The Welding Institute*)

normally associated with EBW in the range of 30in/min (12·5mm/s), it was possible to weld this difficult alloy without cracking even under conditions of fairly severe joint restraint. The reduction in the extent of the residual strain field also allowed successful post-weld heat treatment to be carried out without HAZ cracking.

An example of the benefits of the low shrinkage stresses associated with a rapid thermal cycle occurs with Nimonic 115. This complex heat-resisting nickel alloy is unweldable by conventional fusion techniques because of excessive cracking. The use of EBW allows satisfactory joints to be made up to a thickness of 0·1in. (2·5mm). However, in certain applications, sound deep welds have to be produced in this alloy. It is then beneficial to introduce a shim of Ni–Cr (80:20) material, 0·01in. (0·25mm) thick, between the abutting joint faces before welding. The shim material has the effect of increasing the hot ductility of the weld metal, thus minimizing the tendency to cracking. Even with the use of such techniques, it is difficult to electron-beam weld Nimonic 115 in sections thicker than 0·1in. (2·5mm) without producing microscopic cracks. However, better results might be obtained by an increase of beam power at high welding speeds leading to even narrower weld seams.

When welding refractory alloys of high melting point, the lower welding speeds of arc welding lead to grain growth, as can be seen in Fig. 5.1b which shows a cross-section of a TIG-weld in a tantalum alloy. Figure 5.1a illustrates an EB weld where both the width of the HAZ and the extent of grain growth have been substantially reduced, resulting in a marked improvement in both tensile strength and ductility.

The fusion zone in an EB weld is effectively a fine-grained cast structure, often with directional solidification towards the centre-line of the weld. This defect may lead to cracking, as it represents a line of weakness, although it is possible to minimize this effect by the correct choice of welding speed.

The elementary properties such as tensile strength, ductility, and hardness are affected in a clear-cut way depending on the material type. Whether the solidified weld metal is stronger (harder) than the parent metal depends on whether the material undergoes a hardening phase change on cooling as is the case with most engineering steels. The surrounding heat-affected material undergoes similar heating and cooling cycles although melting is not involved. Since, with

EBW, these zones are confined to a narrow band, the cooling rate is higher than with other processes, resulting in harder structures within the weld. In the parent material, thermal effects, which may result in hardening or softening, are confined to narrow bands. Softening can occur by two basic mechanisms: either through loss of a work-hardened structure or to solution treatment and overageing in a precipitation-hardening material. When welding precipitation-hardening materials with conventional fusion techniques, the weld metal and a zone immediately adjacent to it are softened by dissolution of the precipitate phase at the high temperature to which these regions have been subjected. Further away from the weld metal, there is a zone which is overaged because of the existence of the appropriate ageing temperature for a relatively long period in this region. Full strength or hardness can be recovered in the solution-treated regions by a low-temperature ageing treatment, but properties in the overaged region can be recovered only by a full heat-treatment cycle involving the whole component. This is often impractical and could lead to distortion of the structure. With EBW, the solution-treated areas are narrow and the overaged region is almost undetectable in some materials and is unlikely to have a major effect on joint properties. Because the overaged region is narrow, it is possible to recover almost full properties by the low temperature ageing treatment only, as illustrated by the following test results in an aluminium alloy containing four per cent Cu:

parent material—UTS 27tonf/in^2 (417MN/m^2)
transverse weld—as welded 21–24tonf/in^2 (324–371MN/m^2)
transverse weld—aged after welding 26–27tonf/in^2 (402–417MN/m^2)

A further feature affecting the performance of electron-beam welds is connected with the weld widths and is the relationship between weld width and material thickness. When the depth-to-width ratio of the weld is high, it is likely that the surrounding parent material will play an important part in determining the joint properties. This principle has been demonstrated for brazed joints where the weak braze material is 'supported' by its strong surroundings. With EBW, this principle could be applied for properties such as tensile strength. The position is not clear, however, for properties such as fracture toughness and fatigue strength. In the case of fatigue strength, weld shape has been shown to be the predominant factor even in the presence of internal weld defects. A weld with smooth top and underbeads, coupled with a small angle between weld bead and plate surface, results in good fatigue strength, although this probably becomes less important as plate thickness increases.

Metallurgical limitations of electron-beam welding

Apart from the limitations imposed by the size of the vacuum chamber and the overall cost of the equipment, there are less obvious metallurgical problems associated with the technique. A high-energy beam of electrons is a powerful welding tool with remarkable capabilities, but, equally, incorrect use of EBW can result in a range of weld defects some of which are peculiar to EBW. This range is frighteningly wide at first sight but fortunately nearly all can be overcome by the correct choice of welding parameters or material specification. Many of the defects which can arise are common to all fusion welding processes, for example, cracking in weld metal or HAZ, porosity, lack of fusion, and the production of hard structures. In the following sections, the occurrence of these defects and the ways of controlling them are reviewed.

Porosity

Porosity in the weld metal forms as a result of trapping gas which has been evolved from the liquid metal. The gas can be present in the material to be welded either in solution, as with hydrogen in aluminium, or as free gas pockets in castings. It may form as a result of a chemical reaction during welding as in steels where carbon monoxide is formed as a result of the breakdown of metallic oxides, thus:

$$FeO + C \rightarrow Fe + CO\uparrow$$

Porosity can also arise from surface contamination by grease or oxides which may also contain absorbed water vapour. The necessary preweld treatments vary with the material from simple degreasing to pickling or wire brushing. In some cases, it has been found beneficial to remove chemically a thin layer of the joint surface material before welding to avoid excessive porosity. Figure 5.7a shows a section through a titanium alloy weld which was superficially cleaned before EBW by means of an aluminium oxide abrasive followed by washing in acetone. The incidence of porosity

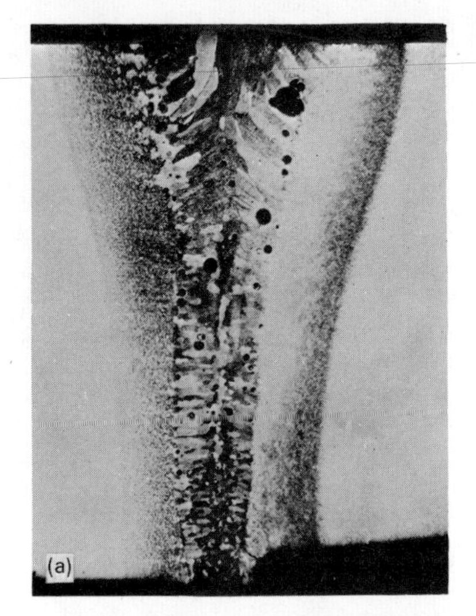

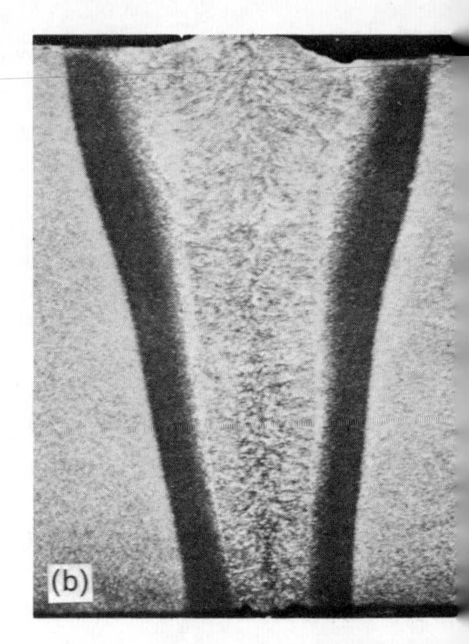

Fig. 5.7 Electron-beam weld of titanium alloy: (a) edge superficially cleaned (× 8), and (b) edge chemically etched (× 11) (*Courtesy Rolls-Royce Ltd*)

is unacceptably high. A similar weld carried out after chemical etching is shown in Fig. 5.7b where it can be seen that porosity has been completely eliminated.

Electron-beam welding of magnesium alloys often gives rise to porosity due to the presence of surface oxides. Cleaning solutions based on chromic oxide have been successfully used. Aluminium alloys which may also suffer from the same defect can be cleaned in solutions based on caustic soda.

The occurrence of porosity as a result of surface contamination is minimized by adequate preweld cleaning, but there remains the problem of gas dissolved in the parent material. In this case, the problem is tackled in two ways. Firstly, by ensuring that the material is degassed to a low level during manufacture, although this is not always economically possible. If the material contains levels of dissolved gas which will give rise to porosity during welding, the second approach must be adopted in which the welding conditions themselves are selected to eliminate or reduce the porosity to an acceptable level. The techniques which are appropriate in this context vary with the material type. In general, for non-ferrous metals such as aluminium and titanium alloys, high welding speeds

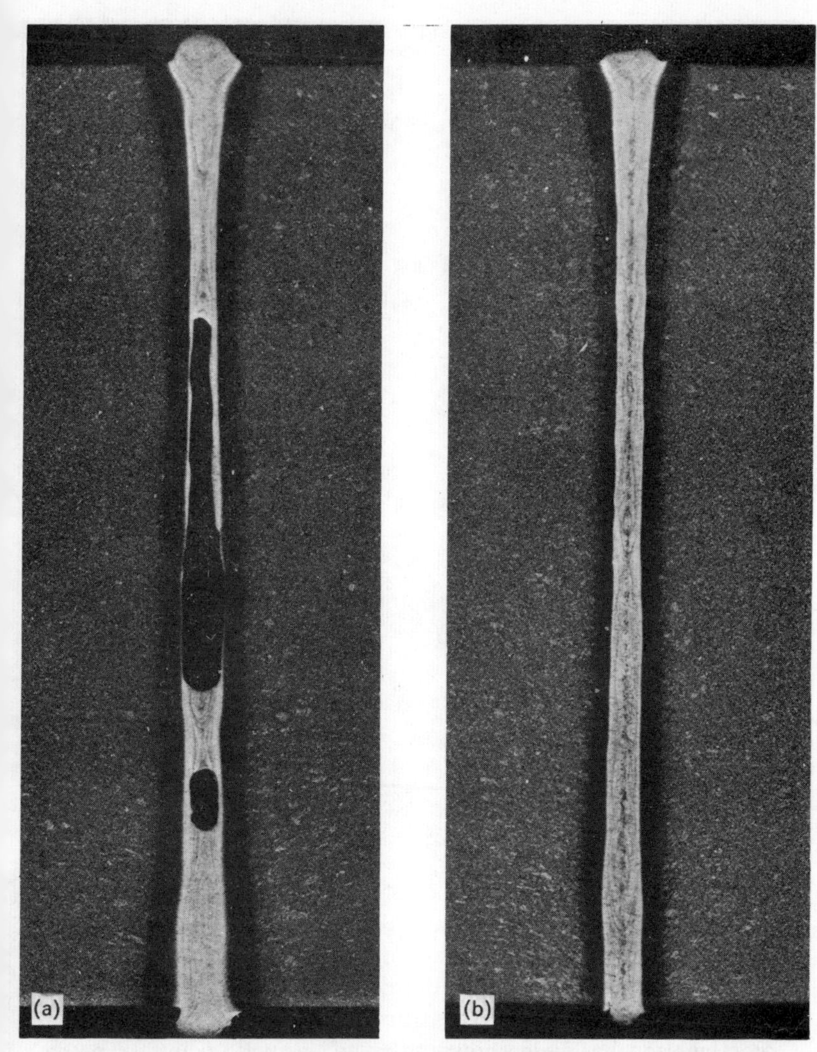

Fig. 5.8 Effect of high-frequency beam rotation in welding a 13 per cent Cr steel: (a) static beam, and (b) with circle generation (× 5) (*Courtesy The Welding Institute*)

of the order of 120in/min (50mm/s) or higher are preferred, whereas lower speeds are beneficial for steels. This situation appears to be associated in the former case with suppression of gas evolution and in the second with the availability of adequate time for vacuum

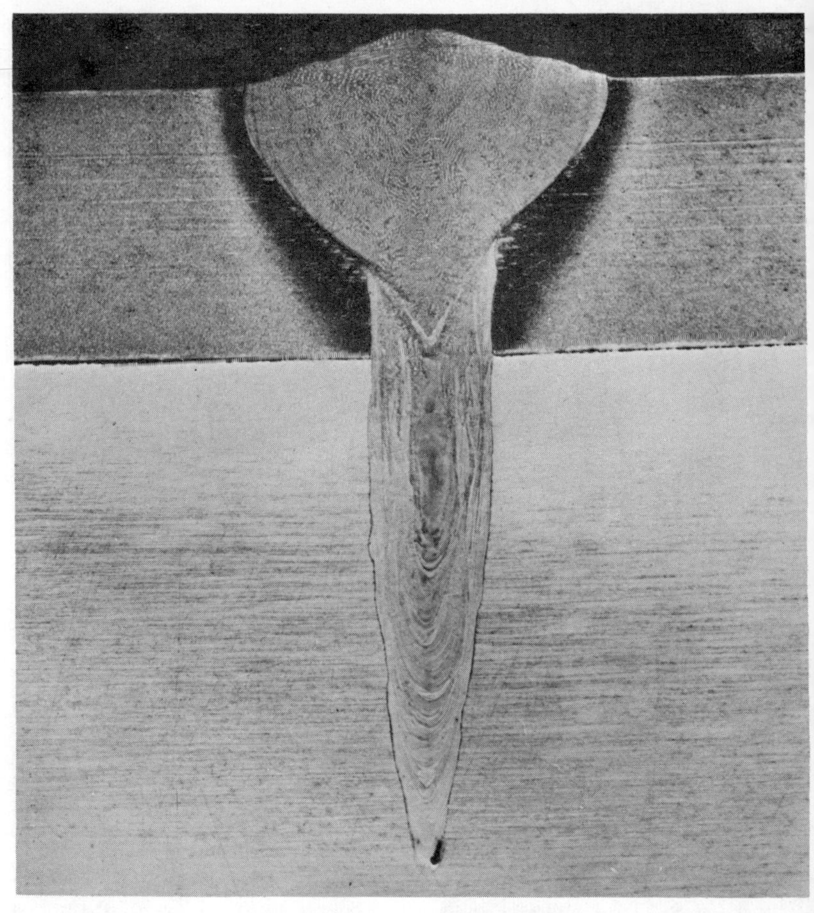

Fig. 5.9 Porosity at root of 'blind' weld (× 20) (*Courtesy Rolls-Royce Ltd*)

degassing of the weld pool. The use of high-frequency beam spinning or oscillation has been shown to be of distinct advantage in the elimination of porosity in steels, permitting steels with high gas contents, such as semi-killed mild steel, to be welded satisfactorily. The use of high frequencies of oscillation or circle generation greater than 500Hz, using relatively small amplitudes, has been found to be much more effective than the application of the standard machine facility of 50Hz beam oscillation. This is illustrated in Fig. 5.8 for a steel containing 13 per cent Cr.

Weld metal porosity from a completely different source arises when attempting to electron-beam weld brass. The zinc constituent has a high vapour pressure and is heated well above its volatilization temperature because of the relatively high melting point of brass. The result is a weld deposit consisting primarily of porous copper since the majority of the zinc is lost from the weld pool.

A 'blind' weld is one which does not penetrate the full thickness of the material; an example being the T weld where the beam penetrates through the horizontal member and partly penetrates the vertical component. Almost invariably, a porous condition will exist at the root of the weld which is thought to result from the entrapment of the penetration vapour cavity, as can be seen in Fig. 5.9. A blind weld should be avoided wherever possible, particularly if the component being welded is subjected to any fluctuating mechanical or thermal stress, since root porosity can constitute a significant stress concentration at the lower surface. In general, wide welds should be used to minimize the occurrence of root porosity since this gives the best chance of metal flow into the cavity.

Hardening

The cooling rate in electron-beam welds is high in relation to other fusion welding processes, because of the local nature of the fusion zone and the high temperature attained in the centre of the weld pool. With steels which are hardenable by a phase change on cooling, structures are produced which are harder than would be obtained with other fusion processes. This is usually due to the formation of a martensitic structure. The hardness reached depends mainly on the carbon content. An as-welded hardness survey across two welds is shown in Fig. 5.10, one produced by electron beam and the other by manual metal-arc. It can be seen that in the fusion zone an increase of some 350 points was found in the electron-beam weld. The HAZ hardness of the arc weld has also increased, but only by some 200 points. In certain materials, welds with a hardness of this order will be prone to quench-cracking under conditions of high restraint, necessitating the use of preweld and/or postweld heat treatment.

The hardness of the arc-weld deposit can be controlled since filler additions are made in the welding cycle, but the HAZ cannot be influenced in this way. With EBW, it is usual to melt only the parent metal although some investigations have been undertaken

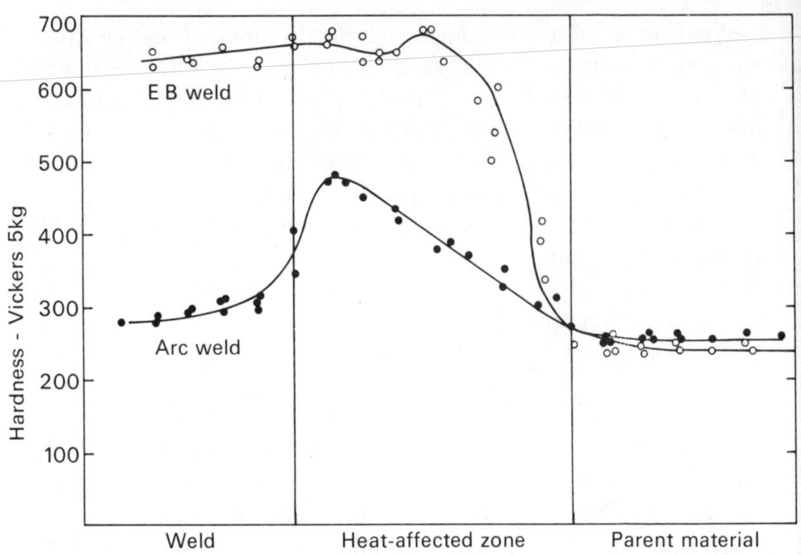

Fig. 5.10 Hardness plots comparing electron-beam weld with arc weld in EN9 (*Courtesy GEC Power Engineering Ltd*)

to assess the value of filler materials in reducing peak hardening. With small components, postweld heat treatment using the electron beam itself as a heat source may be carried out. Such a technique has been used successfully in reducing both HAZ and weld hardness in a 1 per cent C–1 per cent Cr steel from 700 VPN to 550 VPN by inhibiting martensitic transformation and causing direct transformation to a bainitic structure. Since the electron beam is essentially intense and localized, however, it becomes impractical to heat large parts, and some other form of heating inside the vacuum chamber would be preferable.

The high hardnesses of the weld area are generally undesirable because they constitute a sudden change in material properties in the weld zone. In less critical service conditions, however, it is possible to use electron-beam welds in their as-welded state. Some improvement in service performance is obtained in thick sections by the fact that the hardened zone is extremely narrow and is supported to some extent by the softer surroundings.

Cracking in electron-beam welds

Of all possible defects associated with welds made by fusion processes, cracking is probably the most significant from the point of view of service performance. Cracks occur either in the weld metal itself or in the heat-affected parent material. The various types of cracking known to occur in fusion welding are:

solidification cracking;
HAZ burning or liquation cracking;
quench cracking;
heat-treatment cracking;
lamellar tearing.

For a given material composition, the chance of cracking with EBW is reduced in comparison with conventional fusion welding processes, although cracking can still present a problem for many materials and joint designs.

As has been mentioned earlier, cracking is basically a result of insufficient ductility in the material (due either to liquid films or to low-ductility solid phases) to accommodate the contraction strains. The extent of the strain for a given type of weld is a function of the material thickness and joint design, i.e., joint restraint.

Solidification cracking

This type of cracking occurs during the final stages of solidification usually because of the low melting point compounds formed by impurities. An example is shown in Fig. 5.11, where the presence of sulphur and phosphorus in steel is the contributing factor. Low melting point compounds exist at temperatures where the bulk of the material is solid; thus, the grain boundaries have to take large strains. In general, the chance of solidification cracking is reduced if low-restraint joints are used together with low welding speeds. Where economically justifiable, the use of good quality materials with low impurity contents considerably eases the problem, which is most prevalent in engineering steels. In conventional welding, this problem is relieved by the use of good quality filler materials but, with EBW, although filler wire addition can be used, it is difficult to add enough to affect the weld-metal composition significantly; the solution therefore usually lies in choosing the welding parameters which are likely to reduce the risk of this type of cracking. The use of transverse beam oscillation has been claimed to be effective in reducing cracking in a low-alloy steel.[2]

Cracking in the parent material

Although quench cracking, influenced by the presence of hydrogen, is a common problem in the HAZ when welding steels by arc processes, it has never been reported for EBW. This is probably because of the low hydrogen levels and the low strain. The latter

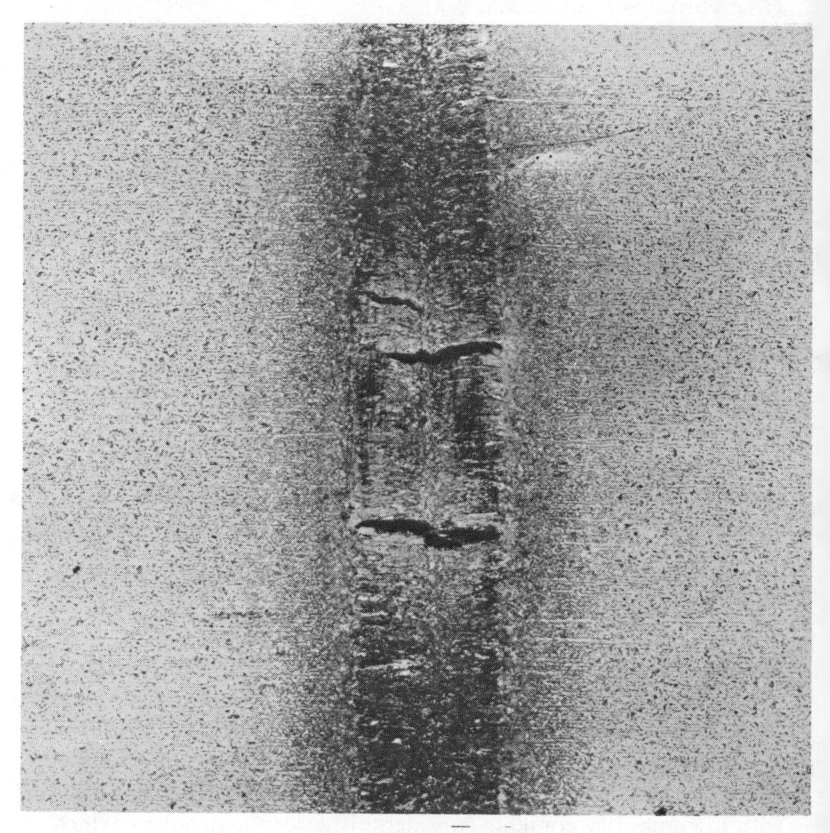

Fig. 5.11 Transverse section showing solidification cracking in a $\frac{1}{2}$in. (12·7mm) mild steel butt weld (× 20) (*Courtesy The Welding Insitute*)

factor probably also accounts for the fact that lamellar tearing has not been reported in electron-beam welds. It also appears likely that HAZ burning in steels due to segregation of impurities by diffusion in the solid phase has not occurred in EBW because of the rapid thermal cycle, although there could be some confusion with another HAZ cracking problem, namely, liquation cracking. This form of cracking is common in nickel alloys (Fig. 5.6) and is due to the presence of liquid films in the HAZ grain boundaries. The formation of liquid films is a function of both the material composition and the weld thermal cycle, and cracking will occur by shear strain in the grain boundaries if sufficient strain is induced in the HAZ. It appears that this type of cracking can be minimized by two mechanisms: a rapid thermal cycle to prevent liquation, or a very slow thermal cycle to provide enough liquid and time for the cracks to heal by liquid flow.

With EBW, cracking in the parent material near the weld has been found in precipitation-hardening nickel-base alloys. Numerous other materials have suffered from the same defect when welded by conventional processes, including austenitic stainless steels and low-alloy Cr–Mo–V steels. Such cracking occurs during heat treatment basically as a result of stiffening of the grains by strain-induced precipitation in material adjacent to the weld during the attempted stress-relief or ageing heat treatment. If the strain cannot be relieved by yielding of the stiffened grains, cracking will occur in the grain boundaries. Because the strain field is extremely narrow with EBW the problem is not widespread and, in cases where it has been found, careful choice of weld parameters to reduce the residual strain field, i.e., to produce narrow welds, has in many instances eliminated the problem.

Quench cracking

Quench cracking can occur during fusion welding in either HAZ or weld metal although in electron-beam welds it has been observed only in the weld metal (Fig. 5.12). This is essentially a brittle crack propagating during cooling, or even after cooling, from a smaller defect which may be a small solidification crack. It occurs under conditions of high restraint involving relatively thick sections in welds with low fracture toughness. This is usually a result of high carbon levels producing very hard martensite in the rapidly cooled weld metal. Cracking is minimized by the use of low-restraint joint designs, but the low fracture toughness of the weld area could well be a problem under some service conditions and it might therefore be preferable to employ pre- and/or post-heat treatment to reduce the hardness. This will prevent cracking during welding and improve the joint properties in service.

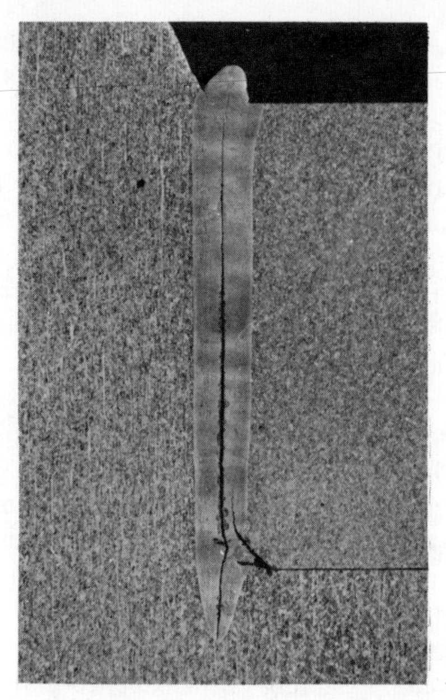

Fig. 5.12 Quench cracking in a 0·4 per cent C steel; weld depth $\frac{1}{2}$in. (12·7mm) ($\times$ 4) (*Courtesy The Welding Institute*)

'Necklace' cracking

This is a form of defect which is peculiar to EBW and has been found in a variety of materials including titanium alloys, stainless steels, nickel alloys, and carbon steels. An example is shown in Fig. 5.13. Examination of the 'fracture' surface and detailed metallographic examination of both transverse and longitudinal sections indicate that the defect is not a true crack but is a type of cold shut. It occurs mainly in blind or very narrow fully penetrating welds. The mechanism of formation is still obscure but it appears to be associated with the inability of the molten metal to flow into the penetration cavity and wet the sidewalls. An alternative explanation is that the stresses associated with the large temperature gradients and cooling rates at the root of a narrow weld are sufficiently high to initiate rupture. Under the influence of contraction and service stresses, the defect can propagate as a true crack. This type of defect can sometimes be eliminated by widening the weld, thus improving the ability of the metal to flow into the cavity and reducing temperature gradients and cooling rates.

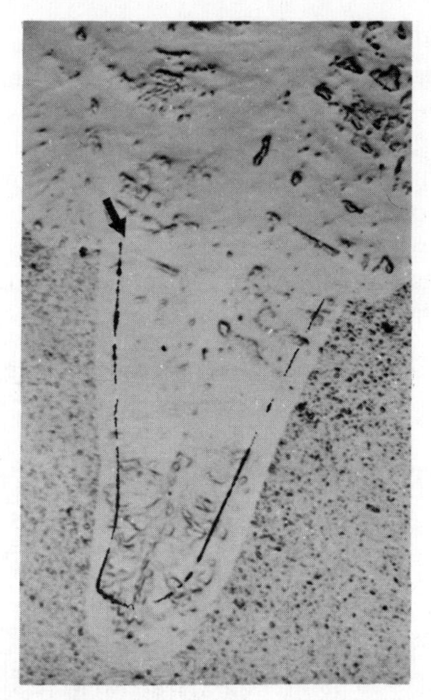

Fig. 5.13 'Necklace' cracking (× 100) (*Courtesy Rolls-Royce Ltd*)

Welding of dissimilar metals

The feasibility of fusion welding two dissimilar metals or alloys is controlled by two main factors. Firstly, there is the physical requirement of melting both metals simultaneously. Naturally, this is more difficult if the two metals have widely differing melting points or coefficients of thermal conductivity. If the mechanism of heating is based on conduction, as in conventional welding methods, it may not then be possible to take even the first step towards welding, namely the fusion of the two metals simultaneously. This is not so with EBW. The mechanism of heating is by exciting the metal atoms, so that, irrespective of differences in melting point or thermal conductivity, both metals are heated and melted when intercepted by the beam. It may be necessary, however, to offset the beam slightly in one direction to overcome large differences in physical properties, but this can be readily achieved.

The second controlling factor in deciding the feasibility of producing a joint between two dissimilar metals is metallurgical. Compatibility is obviously important since, although it may be possible to produce a joint, its mechanical properties may be so poor as to render it useless as a load-bearing member. Naturally, the combination of any two metals or two alloys has to be examined carefully to detect the presence of undesirable brittle intermetallic compounds. With two complex alloys, it may not be possible to reach definite conclusions from a theoretical study of equilibrium diagrams, and some experimental work then becomes necessary. Nevertheless, the chart shown in Fig. 5.14 can be used as a general guide. Areas where undesirable intermetallics are known to exist

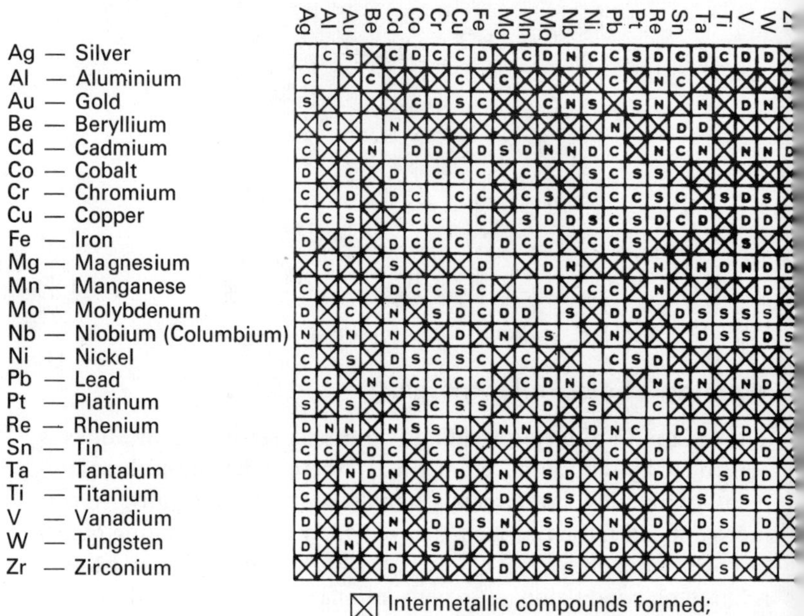

	Ag	Al	Au	Be	Cd	Co	Cr	Cu	Fe	Mg	Mn	Mo	Nb	Ni	Pb	Pt	Re	Sn	Ta	Ti	V	W	Zr
Ag		C	S	X	C	D	C	C	D	X	C	D	N	C	C	S	D	C	D	C	D	D	X
Al	C		X	C	X	X	X	C	X	C	X	X	X	X	C	X	N	C	X	X	X	X	X
Au	S	X		X	X	C	D	S	C	X	X	C	N	S	X	S	N	X	N	X	D	N	X
Be	X	C	X		N	X	X	X	X	X	X	X	X	X	N	X	X	D	D	X	X	X	X
Cd	C	X	X	N		D	D	X	D	S	D	N	N	D	C	X	N	C	N	X	N	N	D
Co	D	X	C	X	D		C	C	C	X	C	X	X	S	C	S	S	X	X	X	X	X	X
Cr	C	X	D	X	D	C		C	C	X	C	S	X	C	C	C	S	C	X	S	D	S	X
Cu	C	C	S	X	X	C	C		C	X	S	D	D	S	C	S	D	C	D	X	D	D	X
Fe	D	X	C	X	D	C	C	C		D	C	C	X	C	C	S	X	X	X	X	S	X	X
Mg	X	C	X	X	S	X	X	X	D		X	D	N	X	X	X	N	X	N	D	N	D	D
Mn	C	X	X	X	D	C	C	S	C	X		D	X	C	C	X	N	X	X	X	X	D	X
Mo	D	X	C	X	N	X	S	D	C	D	D		S	X	D	D	X	D	S	S	S	S	X
Nb	N	X	N	X	N	X	X	D	X	N	X	S		X	N	X	X	X	D	S	S	D	S
Ni	C	X	S	X	D	S	C	S	C	X	C	X	X		C	S	D	X	X	X	X	X	X
Pb	C	C	X	N	C	C	C	C	C	X	C	D	N	C		X	N	C	N	X	N	D	X
Pt	S	X	S	X	X	S	C	S	S	X	X	D	X	S	X		C	X	X	X	X	X	X
Re	D	N	N	X	N	S	S	D	X	N	N	X	X	D	N	C		D	D	X	D	X	X
Sn	C	C	X	D	C	X	C	C	X	X	X	D	X	X	C	X	D		X	X	X	D	X
Ta	D	X	N	D	N	X	X	D	X	N	X	S	D	X	N	X	D	X		S	D	D	X
Ti	C	X	X	X	X	X	S	X	X	D	X	S	S	X	X	X	X	X	S		S	C	S
V	D	X	D	X	N	X	D	D	S	N	X	S	S	X	N	X	D	X	D	S		D	X
W	D	X	N	X	N	X	S	D	X	D	D	S	D	X	D	X	X	D	D	C	D		X
Zr	X	X	X	X	D	X	X	X	X	D	X	X	S	X	X	X	X	X	X	S	X	X	

Ag — Silver
Al — Aluminium
Au — Gold
Be — Beryllium
Cd — Cadmium
Co — Cobalt
Cr — Chromium
Cu — Copper
Fe — Iron
Mg — Magnesium
Mn — Manganese
Mo — Molybdenum
Nb — Niobium (Columbium)
Ni — Nickel
Pb — Lead
Pt — Platinum
Re — Rhenium
Sn — Tin
Ta — Tantalum
Ti — Titanium
V — Vanadium
W — Tungsten
Zr — Zirconium

⊠ Intermetallic compounds formed; undesirable combination
S Solid solubility exists in all alloy combinations; very desirable combination
C Complex structures may exist; probably acceptable combination
D Insufficient data for proper evaluation; use with caution
N No data available; use with extreme caution

Fig. 5.14 Metal combinations for welding dissimilar metals (*Courtesy Hawker Siddeley Dynamics, Electron Beam Division*)

are indicated by a cross. By contrast, areas where solid solubility exists over the whole range of an alloy combination are naturally desirable and are indicated by an S. There are other areas where insufficient data are available.

When joining dissimilar materials, there may be certain advantages in using EBW even though brittle intermetallics may tend to form. Since the weld itself is narrow, the volume of intermetallics may also be reduced to acceptable limits. Again, it may be possible to offset the beam in one direction or the other, thus allowing some control over the composition of the resulting alloy. Although it may be possible to produce a sound joint by these methods on a laboratory scale, it is obviously more difficult to achieve similar control under production conditions. Mixing the molten metal in an EB weld seldom produces a chemically homogeneous fused zone between the two dissimilar materials, even when beam oscillation or spinning is resorted to. Thus, although the average chemical composition of the weld metal may be acceptable, local heterogeneity can be responsible for the presence of brittle zones. It will also be apparent that minor variations in the beam position can significantly influence the relative proportions of the two main constituents in the weld zone.

A technique which is often advocated when joining dissimilar metals by EBW involves the interposition of a shim between the two metals to enhance their compatibility. The thickness of the shim may be critical in controlling the composition of the resulting alloy. As an extension of the principle of a 'bridge', although metals A and B may be completely incompatible, it may be possible to select metal C which is compatible with both A and B. Clearly, the field of welding dissimilar metals is still in its early days of development and a great deal of interesting and potentially useful work remains to be done.

If the two materials to be welded have widely differing thermal expansion coefficients, high internal postweld stresses can be introduced which may lead to cracking. Residual stress measurements made on welds between a steel and a cobalt alloy showed peak stresses of 14tonf/in^2 (216MN/m^2), even after a postweld heat treatment of 600°C for 2h. When concentric rings are welded together and the expansion of the outer ring is considerably greater than the inner, a gap will occur between the two components ahead of the beam which can cause welding difficulties such as overpenetration

Fig. 5.15 Electron-beam weld between Inconel 600 and a high-carbon steel; thickness $\frac{9}{16}$ in. (14·3mm) (×7) (*Courtesy Sciaky Electric Welding Machines Ltd*)

and undercutting. Differential expansion during welding a 6in. (152mm) dia. phosphor-bronze outer ring to a cast-iron inner ring produced progressive relative expansion in front of the beam causing a gap to develop. When the weld was half completed, the gap became so large that the beam passed between the two materials giving no weld at all. By shrinking one ring onto the other before welding, this effect was reduced. Light EB tacking at close intervals around the joint before the full power weld pass can also be beneficial.

The EB welding of Inconel 600, a Ni–Cr–Fe alloy, to a high-carbon steel has been successfully achieved and a macrosection of the joint is shown in Fig. 5.15. This joint, $\frac{9}{16}$in. (14·3mm) thick, was produced without special precautions except that it was found that slow cooling in the vacuum chamber after welding reduced the risk of cracking. The light clamping loads used in the welding fixture were sufficient to keep distortion very small.

A low-alloy steel (0·2 per cent C–0·6 Ni–0·5 Cr–0·2 Mo) has been successfully joined to Hastelloy X, a Ni–Cr–Mo alloy (Fig. 5.16), the welded section being $1\frac{1}{4}$in. (31·8mm) thick. The requirement was to produce a butt weld free from cracking and voids, fully penetrating, and with good weld geometry. Preliminary work indicated certain problems:

machine settings that produced full penetration resulted in severe undercutting and excessive drop-through;
cracking was in evidence;
large internal gas pockets were found.

To prevent excessive drop-through, a backing strip of Hastelloy X was used to support the weld bead while molten; this gave satisfactory weld geometry. Various techniques were then tried in order to overcome the cracking and porosity problems, e.g., placing the backing strip at various distances from the weld, favouring the position of the beam on either the steel or the Hastelloy X. No improvement in weld quality was obtained by these techniques, and therefore it was decided to investigate the use of filler material. It was found that the use of a 0·032in. (0·8mm) shim of 18/8 austenitic stainless steel sandwiched between the steel and the Hastelloy X gave satisfactory crack-free welds. To obtain optimum conditions, a working vacuum of 5×10^{-6} torr was used at the start of the weld and, owing to the outgassing of the Hastelloy X, this pressure increased to 1×10^{-4}torr during welding. The use of the austenitic

Fig. 5.16 Electron-beam weld between Hastelloy X and an alloy steel; thickness $1\frac{1}{4}$in. (31·8mm) ($\times 3\frac{1}{2}$) (*Courtesy Sciaky Electric Welding Machines Ltd*)

stainless steel shim and of the supporting strip, coupled with a good vacuum, produced a satisfactory joint in this difficult material combination.

In some special instances, it is not necessary and may even be harmful to melt both materials. Some dissimilar metal combinations have been achieved by 'braze welding', that is melting one material onto the other using a defocused electron beam as a heat source. Satisfactory joints have been achieved using this technique to join tungsten and molybdenum to stainless steel and tungsten to carbon steel. Success or failure depends on the size of the components and the contraction stresses involved, since brittle intermetallic phases can be formed by diffusion.

Mechanical properties of electron-beam welds

There is little published information on the mechanical properties, particularly the fatigue properties, of electron-beam welds. The majority of the investigations carried out have been confined to tensile strength determinations. Since little purpose would be served by cataloguing a detailed survey of mechanical testing data, some representative alloys will now be examined and general conclusions given. The reader is referred to the published work for further detailed data.

Low-alloy high-strength steels

A typical steel of this type, AISI 4340, has received considerable attention in the United States. Most testing was carried out on specimens welded in the annealed condition and then heat treated.

Published results[3] of mechanical testing indicate that tensile, bend, impact, and fatigue strengths of welds are comparable with the parent material strength.

No problems were encountered in welding this low-alloy material up to $\frac{1}{4}$in. (6·3mm) thick even with no preheat or postweld heating. Conventional arc welding of AISI 4340 necessitates the use of preheat to avoid cracking, and produces welds which exhibit mechanical properties significantly below those of the parent plate.

The EB welding of D6AC material in $\frac{1}{2}$in. (12·7mm) plate has been studied by McHenry, Collins, and Key[2] in which tensile, fatigue, fracture toughness, and stress-corrosion properties were determined. The results of these tests indicate that the welds have mechanical properties that are equal or superior to those produced by TIG-welding. Figure 5.17 shows fatigue S–N curves comparing EB with TIG-welds.

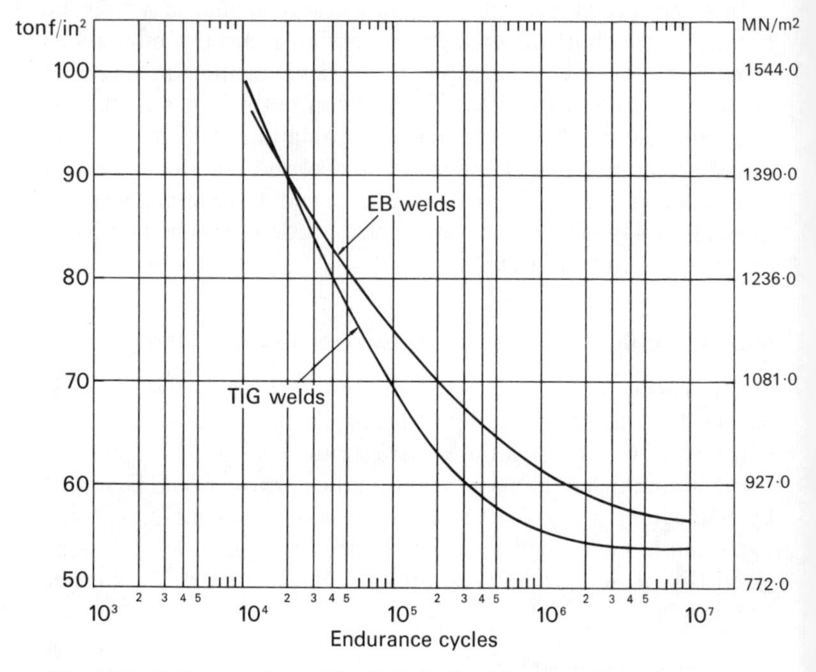

Fig: 5.17 *S–N* curves for welded D6AC alloy (*Courtesy General Dynamics*)

Precipitation hardening high-temperature stainless steels

Electron-beam welds have been made in A286 alloy both in the solution-treated and in the aged conditions. Metallographic examination revealed the presence of microcracks in the HAZ which are typical of the crack susceptibility of this alloy.

The mechanical test results were evaluated by Roth and Bratkovich[4] who showed that up to 90 per cent of the ultimate strength of the parent material can be achieved when the alloy is welded in the solution-treated condition followed by ageing, but only up to 86 per cent when welded in the aged condition followed by re-ageing. Welds made in aged material and tested without further heat treatment exhibited a tensile strength only 66 per cent that of the aged parent material. This is the strength of the solution-treated parent material and the result is to be expected since the weld is essentially a strip of solution-treated metal. White and Bakish[5] welded A286 using a wide beam, 0·03in. (0·75mm) dia., and produced lower strength welds. This reduction in strength is attributed to overageing. The welds made by White and Bakish did not exhibit microcracking and the impact resistance and ductility were high. Roth and Bratkovich carried out bend tests which showed that electron-beam welds give better bend ductility than conventional arc welds using Hastelloy W as the filler metal.

Stainless steels

The properties of electron-beam welds in the EN58 range of stainless steels have been examined by Kenyon.[6] Electron-beam welds on annealed material exhibited tensile strengths with 90 per cent of parent plate strength. This information has been confirmed by one of the authors of this chapter who also found that the fatigue properties of welds are close to those of the parent material. The S–N curves in Fig. 5.18 show a comparison between TIG and EB welds in S521 material.

The tensile strength[7] of welds made in cold-rolled stainless steel, type 301, is approximately 70 per cent of parent material strength. This reduction is due to the heat input on welding which destroys the cold-working properties of the material. All work on this type of stainless steel shows that welding is relatively easy and the weld metal is usually free of defects.

Nickel-based alloys

Meleka and Roberts[8] report that alloys in the Nimonic series which proved difficult to fusion weld by conventional techniques can be satisfactorily joined by EBW. René 41 has been welded by Groves and Gerken[9] who used a variety of preweld and postweld heat treatments. Welds in aged plate followed by direct ageing produced strain-age cracking which is common to arc welds in this material. Optimum results were obtained by means of the following procedure: solution treat, weld, solution treat, and finally age. Transverse tensile strengths approaching those of the aged parent plate were thus obtained. The stress rupture properties of solution-treated and aged welds were found to be inferior to those of the aged parent plate.

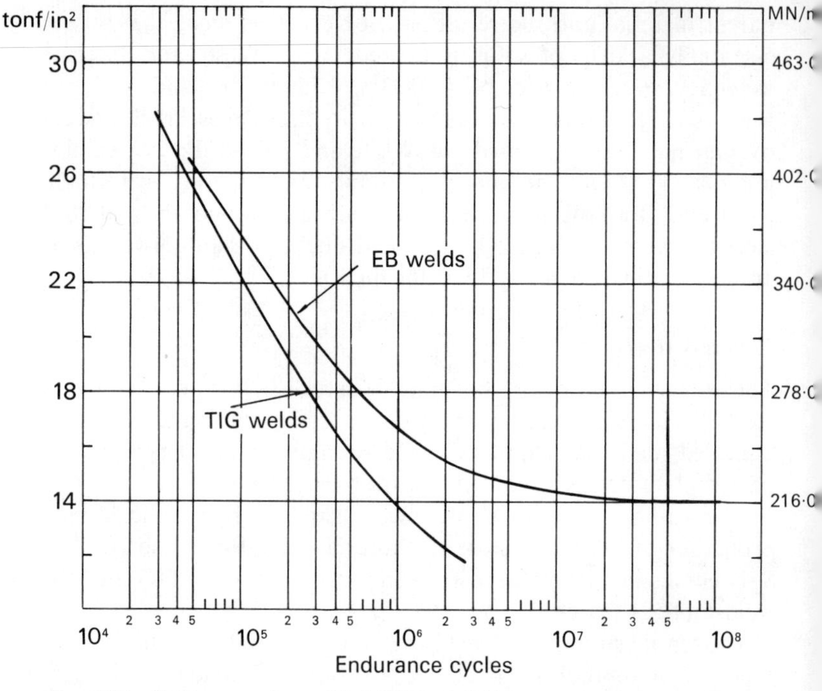

Fig. 5.18 S–N curves for welded S521 steel (*Courtesy Rolls-Royce Ltd*)

One of the authors of this chapter has carried out tests on welds made in solution-treated Nimonic 90. The specimens were solution-treated and aged after welding. The tensile and fatigue strengths of these welds were found to be superior to welds made by arc techniques.

Aluminium alloys

Brennecke[10] reported that 2219 type alloy is readily weldable up to 2·375in. (60mm) thickness. Tensile strengths of 70–80 per cent of the parent material strength as compared with 50–65 per cent with TIG-welds were obtained and the ductility was reported to be good. The presence of porosity in the weld metal was found to have no significant effect on these properties.

Type 6061 alloy has been examined by Meier[3] who reported that a tensile strength about 85 per cent of the parent metal was obtained on welds made in fully hardened material.

Titanium alloys

Groves and Gerken[9] investigated some mechanical properties of joints in a 6 per cent Al–4 per cent V titanium alloy, with various forms of heat treatment. Tensile and stress rupture strengths comparable to parent material strength were obtained; fracture toughness was also reported. The porosity was less than in welds made by arc welding. The alloys C120AV and A70 have been examined by Hokanson and Meier;[11] the tensile strength obtained was equal to that of the parent material.

Welds in a 13 per cent V–11 per cent Cr–3 per cent Al titanium alloy have been tested by Roth and Bratkovich.[4] Parent material strength was sustained both in the as-solution-treated and welded condition and after subsequent duplex ageing.

Beryllium and its alloys

Orner and Hartbower[12] examined the tensile strength of electron-beam welds in beryllium and reported 50 per cent parent material strength. The 68 per cent beryllium/aluminium alloy was welded by Pfluger[13] who observed tensile strength approximately equal to that of pure beryllium.

Molybdenum and tungsten

Sound welds with acceptable mechanical properties were produced in a molybdenum/titanium alloy by Meier.[3] White,[14] examining the same alloy, found the influence of the metallurgical notch to be significant in reducing the strength; widening the weld increased the impact strength. Other work has shown the ductility of the alloy to be low, but the tensile strength is appreciably higher than arc welds in the same material.

Tungsten exhibits a similar metallurgical notch effect to molybdenum, better properties being realized through the use of wider fusion zones.

Conclusions

During the pioneer years of EBW, many unusual metallurgical features and defects peculiar to this new process came to light. However, once the nature of these problems was understood, it was found possible to modify the technique and, on occasion, to control the composition of the welded material more closely so that nearly all these problems have now been overcome.

During the past few years, large quantities of components have been successfully produced by EBW. The performance of these components in service, the increased weld integrity, and in many cases the significant manufacturing cost savings, have amply demonstrated the viability of EBW as a production process.

REFERENCES

1. BAKER, R. G. Anglo-German Welding Conference, Düsseldorf. 4–7 April 1967, 1–11.
2. MCHENRY, H. I., COLLINS, J. C., and KEY, R. E. *Weld. J. Res. Supp.*, **45** (9), 1966, 419s–25s.
3. MEIER, J. W. Second Int'l Vacuum Congress, Washington DC, 1961. Pergamon Press, 1962.
4. ROTH, R. E. and BRATKOVICH, N. F. *Weld J. Res. Supp.*, **41** (5), 1962, 229s–40s.
5. WHITE, S. S. and BAKISH, R. Procs Fourth Symposium Elec. Beam Tech., Boston, Mass., 1962, 496–530.
6. KENYON, D. M. BWRA Report P/17/66, 1966, 20p.
7. BAKISH, R. and WHITE, S. S. *Handbook of Electron Beam Welding*, Wiley & Sons, NY, 1964.
8. MELEKA, A. H. and ROBERTS, J. K. *Brit. Weld. J.*, **15** (1), 1968, 16–20.
9. GROVES, M. T. and GERKEN, J. M. Second Int'l Conference on Elec. Beam and Ion Beam Science & Tech., New York, 1966. Edited by R. Bakish, Wiley & Sons, NY.
10. BRENNECKE, M. W. *Weld. J. Res. Supp.*, **44** (1), 1965, 27s–39s.
11. HOKANSON, H. A. and MEIER, J. W. *Weld. J. Res. Supp.*, **41** (11), 1962, 999s–1009s.
12. ORNER, G. M. and HARTBOWER, C. E. *Weld. J. Res. Supp.*, **40** (10), 1961, 459s–67s.
13. PFLUGER, A. R. Western Metal & Tool Conference, Los Angeles, 1964.
14. WHITE, S. S. *Weld. J. Res. Supp.*, **40** (7), 1961, 317s–19s.

6. Component design

Electron-beam welding is a fusion welding process. In many ways, it must be treated in the same manner as other fusion welding methods: for example, the same metallurgical considerations and inspection techniques would apply. However, certain of its basic characteristics open up such new horizons to the designer that it must be considered as a truly new welding technology.

The purpose of this chapter is to reveal these new possibilities. To secure the true benefits of the process it is essential that the components are designed with EBW in mind. Certainly, the process can often be introduced 'through the back door' so that EBW merely replaces a conventional welding process when the latter has proved to be inadequate or difficult to apply. This is the course of compromise, however; the drawing board is the logical starting point.

Thus, this chapter will attempt to detail what the designer needs to know about the process, which includes not only the potentialities, but also the limitations, which are equally important.

Design capabilities of electron-beam welding

Before discussing the possible impact of EBW on design thinking, the basic design advantages will first be considered.

Very low heat input

It is difficult to envisage a fusion welding method that would require a smaller amount of heat input for joining a given section than EBW. The fusion zone can be so narrow that the least possible material disturbance is produced. In deep penetration welds, the heat input can be as little as one-fiftieth of that encountered in arc-welding processes. Probably the greater benefits are metallurgical but the consequences for the designer can be far reaching.

Distortion of the welded component is the consequence of two distinct actions. There is the obvious thermal distortion resulting from subjecting limited areas of the component to excessive and local heat. The greater the quantity of heat applied, the greater the degree of distortion. There is also distortion resulting from the displacement of component parts owing to shrinkage of the molten metal upon solidification. The designer aiming at an integrated component of specified dimensions has to allow for the anticipated weld shrinkage. Again, shrinkage in EBW is the smallest that can be expected by fusion welding. For example, when welding a steel section 0·500in. (12·7mm) thick, a shrinkage allowance of 0·004in. (0·1mm) is required for EBW, compared with 0·020in. (0·5mm) in arc welding.

Not only is the shrinkage distortion smaller in EBW, it is also more uniform. This is because the fusion zone is almost parallel-sided compared with the almost semi-circular geometry of arc welding. The wider upper zones of arc welds thus tend to produce additional displacement of a rotational nature as illustrated in Fig. 6.1.

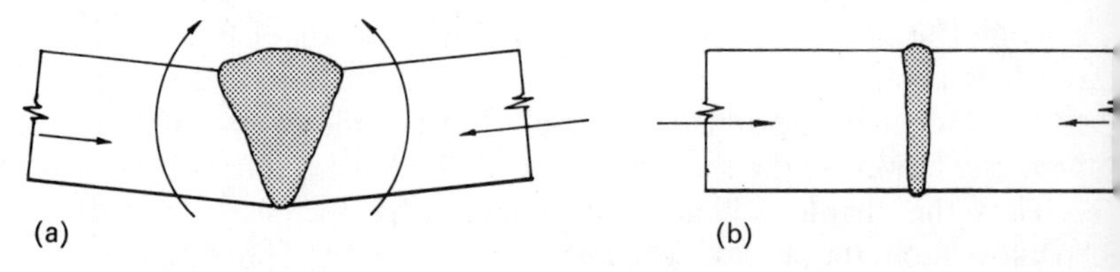

Fig. 6.1 Shrinkage distortion: (a) arc weld showing lateral and angular displacement, and (b) EB weld showing only small lateral displacement

There are undoubtedly advantages to be derived in the utilization of lighter, simpler, and hence cheaper fixtures during welding by electron beam. It is possible to envisage cases where no fixtures are required: only a few tack welds would suffice.

There are other benefits to be derived from the low heat input characteristics of EBW. The component parts can be finish-machined before integrating by welding, thus reducing the cost of machining considerably. Where the thermal distortion by conventional welding is unacceptably high, machining of the integrated part becomes essential. Because the part is then larger or the machining operations intricate, machining costs are often high.

The magnitude of the internal stresses in a welded component is naturally related to the amount of heat induced during welding. Often, the relaxation of the component upon heat treatment results in cracking in the weld or the HAZ. This tendency is also reduced, or even eliminated, when EBW is employed. Internal stresses may produce cracking after the component has been stressed in service. This is naturally a more serious situation, but again EBW reduces such risks.

Welding 'unweldable' materials

A material is considered unweldable mainly because it cracks either during welding or later after heat treatment. Cracking often results from the strain induced by shrinkage which may exceed the fracture strength of the hot HAZ or the solidifying material. Since the shrinkage strains are quite small in EBW, certain materials which normally crack when welded by conventional methods may survive when welded by an electron beam. It is important, therefore, for the designer to rewrite his weldable alloys list to incorporate those materials that can now be successfully handled by electron beam. This aspect is covered in greater detail in chapter 5.

Long focus

The focal length of an electron lens can be quite large. The electron-optics considerations have been dealt with in some detail in chapter 2; it is only necessary to say here that, in commercially available EBW machines, the distance between the magnetic focusing lens and the plane of welding can be as long as 4ft (1·2m) in high-voltage equipment. It is no longer necessary, therefore, to place the heat source immediately above the component surface, as is certainly the case in an electric arc, a gas flame, or even a plasma jet. Thus, the designer can specify welds in narrow and restricted areas where the geometry would have made it impossible for a welding torch to be placed. This is illustrated in Fig. 6.2 where a deep channel and a spool are produced by welding two halves.

The electron beam is also extremely fine: it can pass through narrow gaps as small as 0·025in. (0·6mm) although care must be taken in such extreme cases. This capability undoubtedly helps to reach the more inaccessible areas in components with complex or fine geometries. It is also possible to produce welds close to the surface of a component part. Thus, for example, the rib shown in

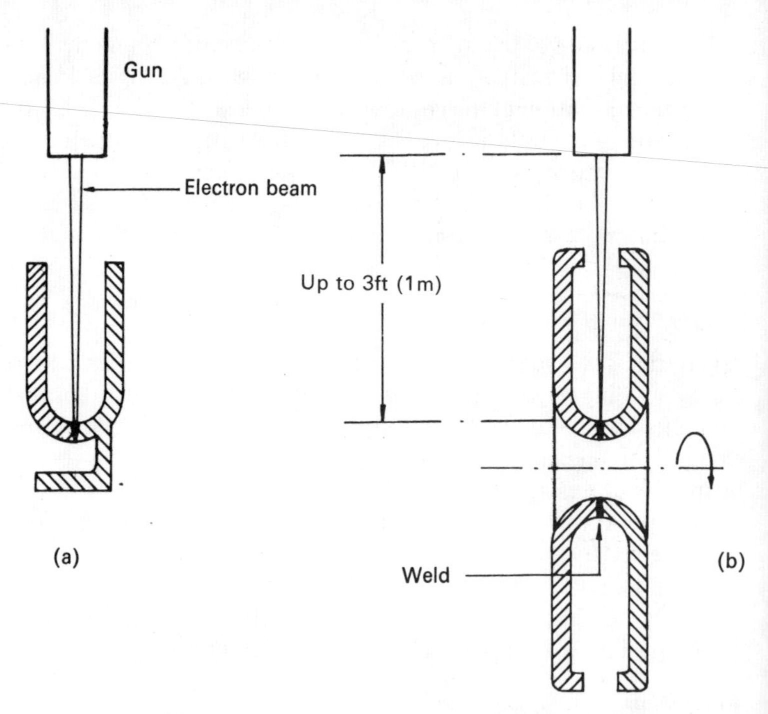

Fig. 6.2 Possible applications of EB, long working distance: (a) channel weld, and (b) spool weld

Fig. 6.3 can be made quite small. This reduces the cost of machining and wasted material. Sometimes the ribs may be produced by chemical milling, often a cheaper and more convenient way of producing this type of configuration. The rib could be dispensed with altogether, although here it is naturally necessary to orientate the beam at a small angle to the surface.

Wide range of thicknesses

The capabilities of deep penetration by electron beam should not detract from its usefulness in welding very thin sections. By adjusting the welding parameters, the same EBW machine can weld sections only a few thousandths of an inch thick as well as metal plate a few inches thick (0·1–100mm). This wide spectrum must be appreciated by the designer, since sections that have been hitherto either too thick or too thin can now be welded, and by the same machine.

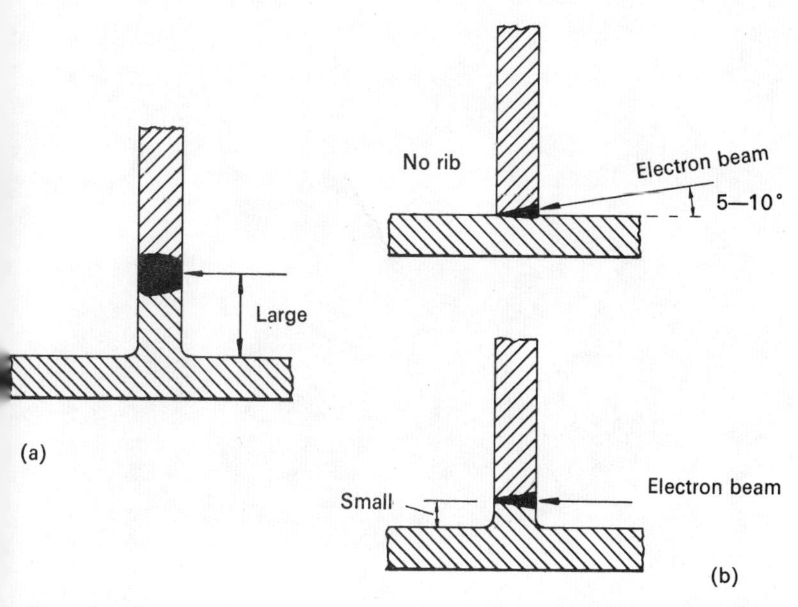

Fig. 6.3 Welds produced close to metal surface: (a) arc weld, and (b) EB welds

It is also possible to weld thick to thin. Conventional methods preclude such a possibility since heat is transmitted to the metal by conduction and the quantity of heat required to fuse the thin section is not adequate for fusing the thick section, but if the heat source is intensified to meet the need of the thick section, the thin component merely melts away. By contrast, with EBW, heat is generated within the material when it intercepts the electron beam, whatever the thickness. True, more heat may be carried away by conduction in the thicker section but this may be overcome by offsetting the beam slightly towards the thicker section. A remarkable demonstration is shown in Fig. 6.4.

Welding dissimilar metals

The metallurgical aspects of welding dissimilar metals were dealt with in the previous chapter. What is worth noting here is that it is now possible to design components in sections made up of different materials to meet specific local requirements, instead of making the whole part of one material and so accepting penalties of cost or weight, etc. The torque-transmitting part of a shaft can be made of

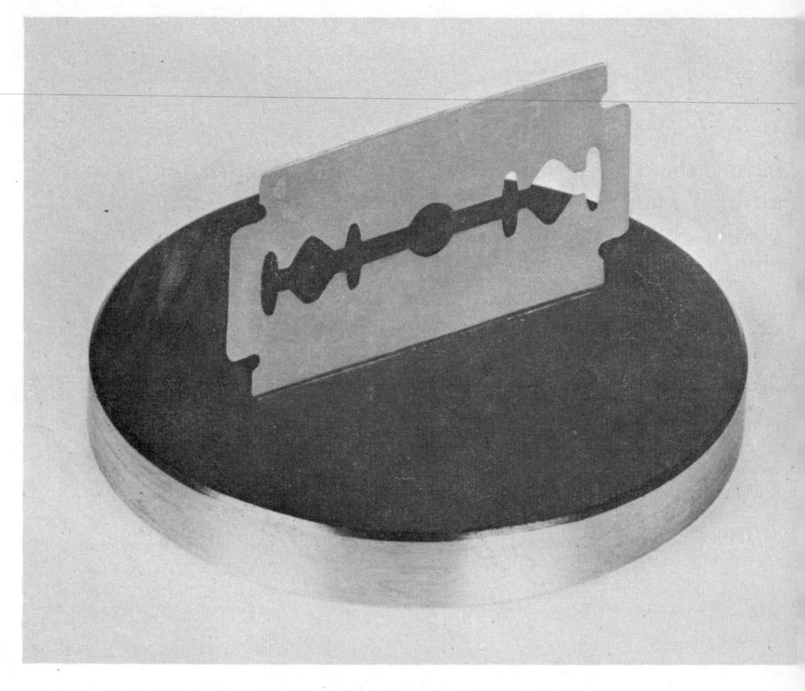

Fig. 6.4 Welding thick to thin (*Courtesy Carl Zeiss*)

one material and the wear-resisting section of another. In electrical machinery, electrically conducting copper sections of a component may be welded to a steel section for increased mechanical strength, and so on.

High welding speed

Generally, welding speeds by electron beam are some ten times faster than electric-arc methods when the penetration is achieved in a single pass in both cases. Naturally, with greater beam power, higher welding speeds can be achieved. Further, when many runs are necessary in arc welding methods, EBW, with only one run, can complete the weld in far less time. This aspect must be taken into account when considering the economics of comparison. The EBW plant is certainly many times more expensive than its arc counterpart, but it can handle work at a considerably faster rate even allowing for the pump-down time of the vacuum chamber. Labour costs can indeed be lower when EBW is employed.

Accuracy of repeatability

Excellent welds can be produced by arc welding, and mechanization has reduced the reliance on operator skill. In EBW, the process *has* to be completely mechanized because the welding speeds are well beyond the response abilities of the operator. Further, the design of an EBW plant lends itself quite readily to automatic control. It is possible, therefore, to obtain a high degree of accuracy of repeatability. The process is such that the weld is generally either good or bad; the 'grey area' is almost non-existent. This reduces the burden of inspection and increases the reliability of the process.

Multiple penetration

A feature of EBW is that it makes no difference to the beam whether it penetrates a thick plate or a number of thinner layers whose total thickness is equal to that of the thick plate (Fig. 6.5). In other words, if the beam power is capable of penetrating thickness t in a given material at a given welding speed, and if the beam is intercepted by a layer of the same metal of thickness $t/2$, the excess power leaving the lower face of this layer is sufficient to penetrate a second layer of thickness $t/2$. This feature is unique to EBW and opens up remarkable new design possibilities.

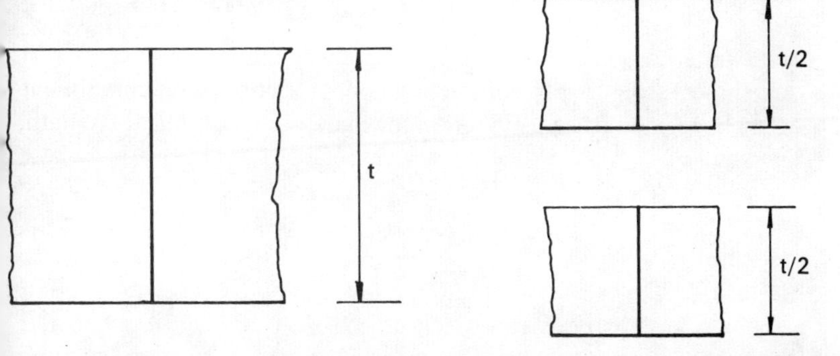

Fig. 6.5 Equivalent multiple penetration

However, there are one or two aspects of this characteristic that have to be closely considered before applying it. It is clear that the welding of the lower layers will be carried out 'blind'; only the top layer will be visible to the operator. It follows that the machining accuracy of the abutment surfaces has to be of a high order. Further, and this might be overlooked, the component has to be accurately set up so that all abutment lines are aligned in the same plane of the beam.

It is clear that the top layer in a multiple-penetration welding operation is subjected to a beam power well in excess of the minimum required for penetrating that layer alone. Certain alloys are less sensitive to such a condition, while others produce channelling on the top bead and 'drop-through' on the lower bead. Naturally the effect is reduced at the lower layers.

Multiple penetration is likely to find application in such fields as heat exchangers and rib-stiffened structures, as illustrated in Fig. 6.6.

Fig. 6.6 Application of multiple penetration to a rib-stiffened structure (*Courtesy Carl Zeiss*)

Limitations of electron-beam welding

It is important for the designer to be equally as aware of the limitations of a new technique as of its capabilities. However, some limitations can often be overcome by further development whereas others may be so basic that they will have to be accepted for a long time to come. This distinction will be made in the following pages and the designer is advised to consult the welding engineer, as new developments are likely to emerge from time to time.

The vacuum limitation

The dimensions of the vacuum chamber impose an upper limit on the size of component to be handled, since the practice now is to place the whole component in the vacuum chamber. However, techniques are being developed where only the area to be welded is surrounded by the vacuum environment. Thus two suction 'cups' are placed one above and one below the abutment edges. The lower 'cup' is essential since, without it, air would leak through the abutment during evacuation before welding and, even if this difficulty were overcome, the liquid metal formed during welding would be pushed back by atmospheric pressure into the upper vacuum section containing the gun. Local vacuum chambers, shown diagrammatically in Fig. 6.7, have been utilized in welding aircraft wings to the fuselage.

It is worth noting, however, that very large chambers have already been constructed, sometimes large enough to house a 10ton truck or a satellite-carrying rocket.

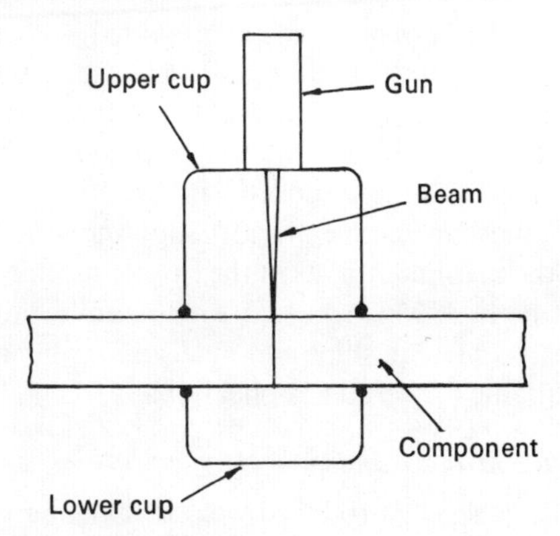

Fig. 6.7 'Local' vacuum chamber

Besides the physical limitations of the dimensions of the chamber, there is also the fact that EBW plant is generally heavy and hence immobile, so that the work has to be brought to the machine.

Again, there is the additional pump-down time of the chamber; the larger it is, the longer the pump-down time. However, this factor has to be weighed against the very high welding speeds achieved by EBW. It seems likely that pump-down time will be considerably reduced by the application of soft-vacuum methods as described in chapter 4. One or two minutes will be the normal time as compared with the 10–15min now required with hard vacuum.

Magnetic materials

An electron beam is readily deflected by small magnetic fields. Thus, welding of materials with retained magnetism presents special problems since the alignment of the beam is likely to be disturbed by the component's own magnetic field. This does not necessarily exclude the welding of magnetized materials by electron beam; it may be possible to compensate for magnetic deflection by offsetting the beam by the necessary amount, although this practice is rather difficult to apply. The fact that the magnetism of certain alloys is often altered by the heat applied during welding and by the consequent metallurgical transformations may also cause difficulties. It is now the practice to demagnetize all ferromagnetic materials before welding.

Profile welding

The majority of EBW machines are provided with manipulators capable of producing linear and circular welds. Such welds represent by far the largest proportion of all welds produced in industry, but there will be occasions when the design demands a profile weld. Because of the high welding speeds of EBW, profile welding has to be carried out by mechanized means, although numerical control can be applied to the manipulator, as in the case of machine tools. There are also other methods which make use of line-following devices, e.g., flame cutting steel plate in shipbuilding. Cams and template-followers are also suitable for certain applications.

Variable working distance and variable penetration

By far the most generally produced weld is that where the penetration is constant and the upper surface of the abutment is presented at a constant distance from the electron gun. Under such conditions, both beam power and the focusing current in the magnetic lens remain unaltered during the welding operation.

Naturally, there will be times when either the working distance varies because of the complex geometry of the component, or when the thickness to be welded varies along the length of the weld line. This may present difficulties to the welding engineer who will have to resort to costly fixtures that will present the component surface at the same level to the beam or, possibly, to some form of automatic control of beam focus. For variable penetration, the simplest answer is to maintain the beam power constant but to vary the welding speed to suit changes in section.

It should be pointed out here that working height tolerances can be surprisingly high in certain cases, up to 2in. (50mm). This depends, however, on the gauge of the material to be welded and its physical properties, particularly viscosity and surface tension around its melting point. Also, it is possible with the same beam power and welding speed to penetrate a section whose thickness varies by up to 30 per cent, but each case will have to be judged on its own merits.

Accessibility

It has already been pointed out that the slender path of an electron beam can find its way through narrow gaps and can reach 'inaccessible' regions. However, areas to be reached by the electron beam must be 'seen' by the gun; in other words the path of the electrons from the gun to the area to be welded is a straight line. In conventional welding methods, the manual weld can be produced in difficult areas, provided there is enough space to place the torch close to the weld line.

This is a basic limitation of EBW and there is little that can be done to overcome it, at least in the near future. It may be possible to visualize a flexible gun in which any mechanical flexing would be accompanied by a corresponding magnetic deflection of the beam, but such a device remains as an interesting concept for the moment.

Criticality of machining and assembly

The heat produced by a well-focused beam is extremely localized; this is the main asset of the process. It is, however, not without its penalties. The abutment faces have to be machined to close tolerances since a gap above, say, 0·005in. (0·1mm) cannot be tolerated.

This naturally adds to the cost of preparation. It also largely excludes sheet metal applications where such accuracies are difficult to achieve. The wider application of automatic wire-feed devices will ease the situation but the basic requirement will continue. It must be remembered, however, that the quality of the product is much improved by the precision requirements of the process, and the additional cost may be fully justified.

It is not sufficient for the butting edges to be accurately machined; the component parts will have to be accurately fixtured before welding. Figure 6.8 illustrates a comparison between an arc weld and an electron-beam weld in which the parts are slightly displaced vertically. Because of the greater size of the arc weld, the step is smoothed over and can possibly be tolerated, but this is not so in the case of an electron-beam weld.

Fig. 6.8 Importance of alignment in (a) arc weld and (b) EB weld

Considerations of cost

Many users have taken the view that the process should not be introduced merely because it is new; it should be considered only where more conventional methods have proved to be inadequate by virtue of a novel design concept or the introduction of a new material. This is obviously wise, since the plant is expensive. Thus, applications have to be selected with care and when this is done it can often be shown that the high capital cost is more than offset by the excellence of the new product or even in the reduction of the rate of rejection or scrap.

Cost considerations, as they affect design, deserve special attention and will be treated in a separate section of this chapter. Cost analysis will be examined in chapter 8.

Impact of electron-beam welding on design

Let us now attempt to sum up the impact of EBW on component design. Perhaps the most important consideration is that of attitudes. Because of the frequent reliance on the skill of the welder when conventional methods are used and the inevitable variability in the quality of the welds produced, there is in many industries understandable prejudice against welding generally. Such attitudes will have to be re-examined in the light of the new capabilities of the electron beam and, where appropriate, such prejudices should be discarded. This process has to be tempered with caution and it should not be assumed that EBW is the answer to all welding problems. In certain industries, fabrications are not normally considered, probably because the thickness of the components handled by that industry has been beyond the capabilities of conventional techniques; or, at the other end of the spectrum, the parts are too fine or the operation too precise for conventional welding; or again, no thought was ever given to the practicability of welding thin sections to thick ones or to the joining of dissimilar materials. It is clear that design ideas will have to be re-examined and practical evaluation undertaken. It is not essential to acquire plant at such an early stage; rather, use can be made of existing facilities in one of the jobbing shops in existence in this country. Such organizations have acquired a substantial amount of know-how derived from the varied nature of the tasks undertaken over the years. Valuable expert advice can also be obtained from The Welding Institute, which has devoted substantial research resources over the last few years to a study of the process.

Joint design

It is often possible to produce a given joint configuration by more than one joint design. The selection is naturally arrived at by considering the various competing factors such as accessibility for welding and inspection, cost of joint preparation, and both the nature and level of the stress to which the joint is subjected. It is true that EBW offers a greater freedom to the designer by virtue of the long working distance, the ability of the beam to find its way to the joint through narrow gaps, and, because of the localized nature of the heat spot, the possibility of welding in areas close to finish-machined surfaces.

The following are illustrations of joint configurations that can be readily produced by EBW. In each case, advantages and disadvantages are outlined and comparisons made with other alternatives.

Butt joint

A butt joint is one of the most widely used joints in welding and deserves special attention. Also, many of the basic features revealed by a close examination of a butt joint are relevant to other types. It is the joint generally recommended for EB fabrications.

It is a general characteristic of an electron-beam weld that the weld zone narrows down from a wide upper bead to virtually a line joint at the lower bead, as shown in Fig. 6.9a. This configuration represents the minimum heat input, an obviously desirable feature. However, there are a number of potential hazards that have to be watched closely. An excessively narrow fusion zone often harbours root porosity as shown in Fig. 6.9b. This is probably because the metal vapour generated at the centre of the molten column is unable to escape through the narrow slot at the bottom of the joint interface. Upon condensation and solidification, a void is formed.

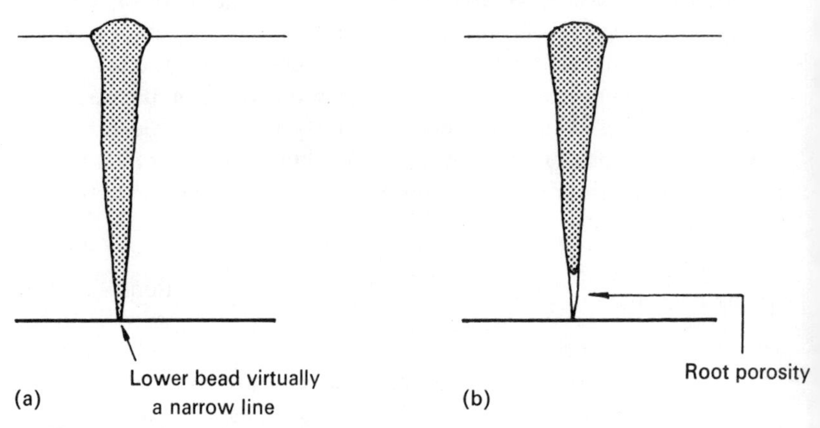

Fig. 6.9 Classical 'dagger' weld (a) showing root porosity (b)

Particularly in deep-penetration applications, it may not be possible to achieve fusion of the whole depth of the interface if the fusion zone narrows down to a 'dagger' edge, as shown in Fig. 6.10a. This is a serious occurrence and is sometimes particularly difficult to detect. It must also be appreciated that small variations in welding parameters may lead to local lack of full penetration, as in Fig. 6.10b. Thus, at one point along the weld line, A, full penetration is achieved, but at B a no-weld condition exists at the root. Because of the smallness of the apex angle of the triangles representing the volume

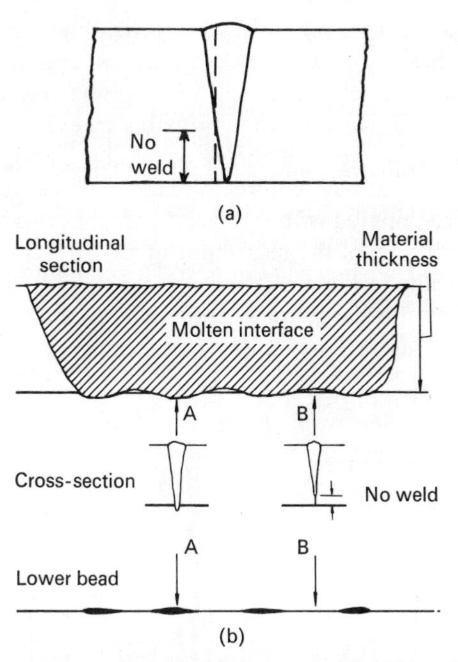

Fig. 6.10 Incomplete penetration of narrow weld owing to (a) alignment difficulties, and (b) small local variations in weld settings

of molten metal, even small local variations in heat input may produce appreciable differences in weld penetration. It is comparatively easy to establish whether such an erratic penetration situation exists by examining the lower seam, where a 'beady' appearance, as illustrated in Fig. 6.11, is normally in evidence.

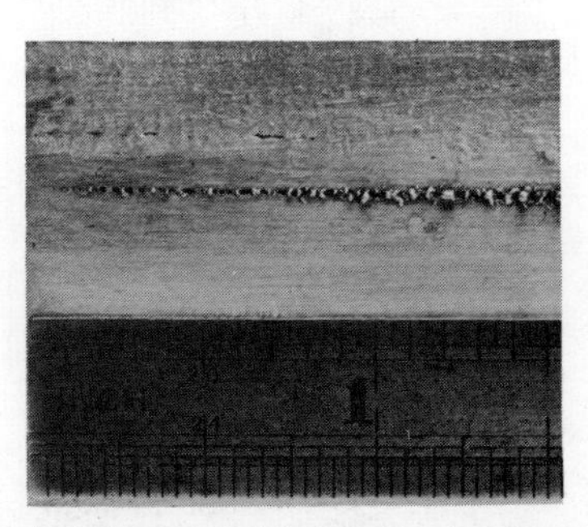

Fig. 6.11 Undesirable 'dagger' weld showing beady appearance of lower seam (*Courtesy Rolls-Royce Ltd*)

It is advisable, therefore, to select the welding parameters which will form a 'healthy' lower bead. It must be appreciated that the heat input per unit length of weld will then be greater than in the truly narrow weld. However, it is probably true that all the difficulties outlined above are then largely overcome. The EB weld with a healthy lower bead may require, for example, twice the amount of heat input as compared with a 'dagger' weld, but this is nevertheless only one-tenth, say, of the heat input of an arc weld, as illustrated diagrammatically in Fig. 6.12.

An unacceptably large underbead may result in some molten metal drop-through and channelling on the upper surface (Fig. 6.13):

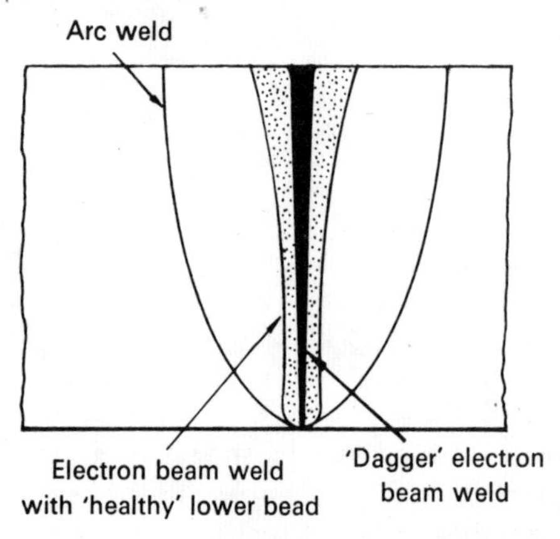

Fig. 6.12 Comparison of volume of fusion zone

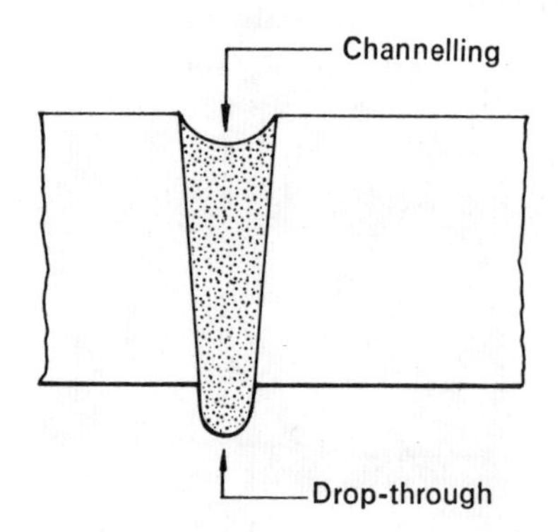

Fig. 6.13 Undesirable features of excessively wide EB welds

both are undesirable features because of their stress-concentration effects. Some surface machining would indeed provide the answer, if it is feasible, but machining after welding is an added operation extending both the manufacturing cycle and the production cost. It is often possible and certainly worth while to attempt to produce a weld with acceptable upper and lower bead geometry.

A 'lip' preparation often used in butt welding is illustrated in Fig. 6.14 and provides the necessary additional material for the formation of satisfactory upper and lower beads.

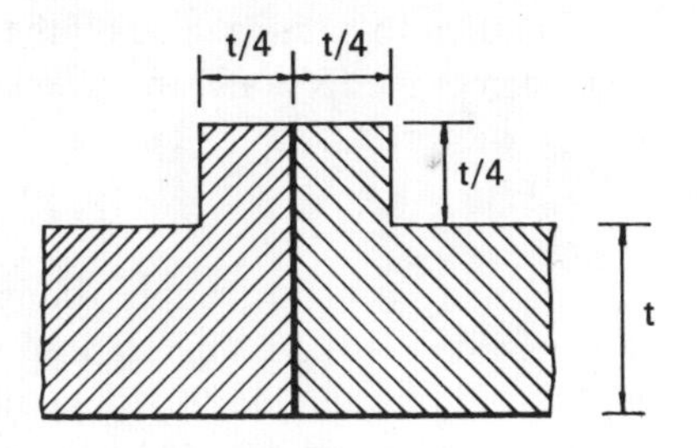

Fig. 6.14 Geometry of 'lip' preparation

A variation on this geometry is shown in Fig. 6.15 which has the benefit of relative location of the two parts. However, the weld line is then invisible and a marker has to be accurately scribed as indicated. The control of the lower bead geometry is dependent on the physical properties of the molten material; the optimum shape can be achieved only by experimentation.

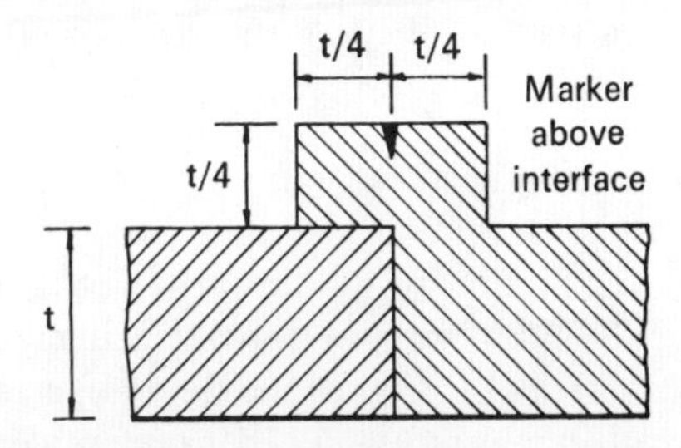

Fig. 6.15 Alternative 'lip' preparation providing location of parts

Integral or separate backing support to the lower bead may be provided (Fig. 6.16). A disadvantage is that the lower bead cannot be inspected and, since the penetration may not be complete, root

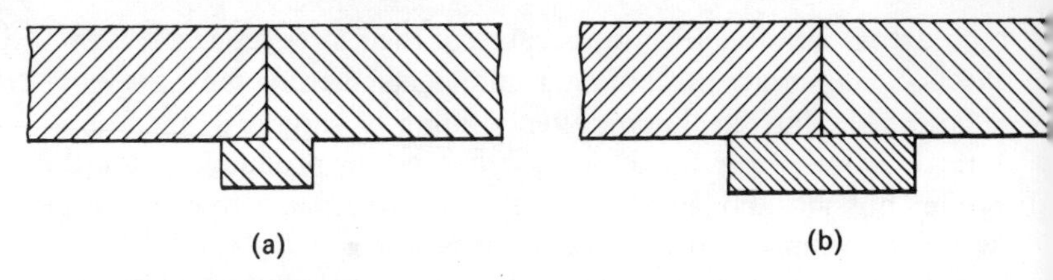

Fig. 6.16 Backing support to lower bead: (a) as integral part of the joint design, or (b) by means of a separate plate

porosity may exist. It is advisable to machine off the extra material, thus removing any such imperfections and revealing the lower sections of the weld for inspection.

A T joint

A T joint is popular in many structural applications because of the stiffening support it provides. It is possible to produce such a joint by EBW by 'nailing' a weld through the horizontal member into the thickness of the vertical one, as can be seen in Fig. 6.17a. This may

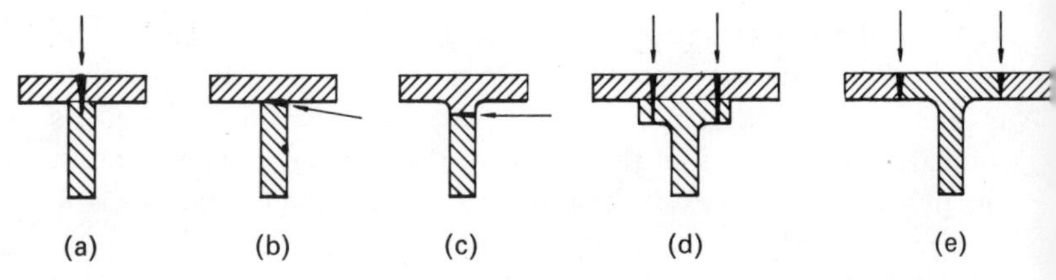

Fig. 6.17 Various designs of T joint showing progressively improving integrity (a–e)

be acceptable for non-structural joints or even for lightly stressed ones, but such an arrangement has many serious drawbacks. The joint line is not visible during welding; the lower bead cannot be inspected; being a blind weld, root porosity is likely; and the whole joint is held by a narrow section of the weld nugget. If accessibility allows it, the arrangement in Fig. 6.17b is preferable. The beam will have to approach the joint line at a small angle to the surface and the weld must be wide enough to allow for such an approach angle. It can be seen that the objections associated with the first arrangement have all been overcome, but better still is the arrangement shown in Fig. 6.17c, mainly because of the reduced stress-concentration effect.

A T joint can be achieved by means of two lap welds as shown in Fig. 6.17d, but perhaps the soundest design is that illustrated in Fig. 6.17e which relies on two butt joints.

Flange-to-shaft joint

A flange-to-shaft joint is frequently found in hydraulic and mechanical drive applications. Perhaps the simplest form is to cut a hole in the flange or plate equal to the outside diameter of the shaft or pipe. A planetary butt weld is then produced as can be seen in Fig. 6.18a. The beam will have to approach the pipe surface at a small angle if the joint line is inaccessible from the other side of the flange. An improvement is shown in Fig. 6.18b which reduces the geometrical stress concentration. However, both arrangements are based on a

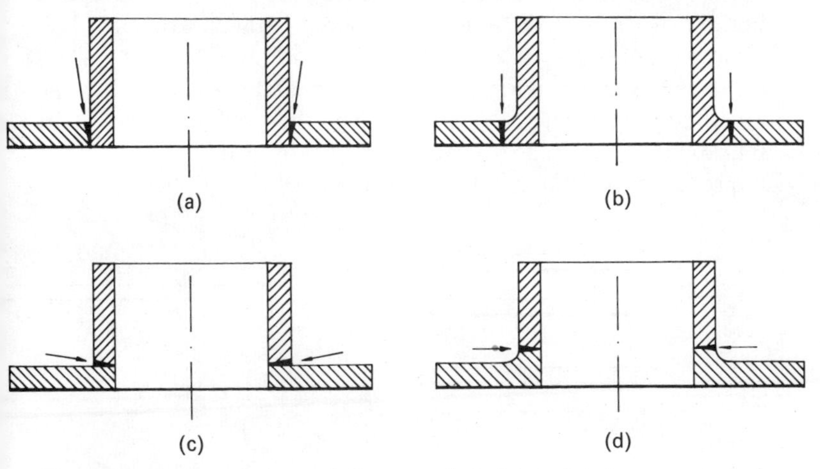

Fig. 6.18 Various designs of flange-to-shaft joint showing progressively improving weld (a–d)

planetary weld which is difficult because of the progressive shrinkage that takes place as the weld proceeds, making it difficult in extreme cases to secure a gap-free abutment towards the end of the weld. Even if this difficulty is overcome, the joint produced is often a stressed one, which in extreme cases may result in cracking during welding, during heat treatment, or in service. It is better to employ a circumferential weld as shown in Fig. 6.18c. The diameter of the hole in the plate is now equal to the inside diameter of the shaft. The joint is easy to produce and is stress-free since the component parts are free to move in the direction of the shaft axis to accommodate shrinkage displacements. Such a movement can be allowed for when the two parts are machined before welding so as to achieve the final dimensional accuracy of the finish-welded part. A local increase in the flange thickness (Fig. 6.18d) improves the accessibility for welding and reduces the stress concentration in the weld zone.

Lap joint

Generally, a lap joint should be avoided whenever possible and replaced by a butt joint. A lap joint, however, is easier and cheaper to prepare and in certain applications may be quite acceptable. The simplest form of an electron-beam welded lap joint is illustrated in Figs. 6.19a and b. In both cases, the load-bearing section is only a narrow zone in the weld nugget. Figures 6.19c and d represent an improvement since the load-bearing area is much extended. It is preferable, of course, to convert the joint in Figs. 6.19a and c into a butt joint, as shown in Fig. 6.19e. If the ends of the two plates in b and d can be bent slightly, this will assist in the welding operation by improving accessibility to the beam and by producing a slightly improved stress system in the joint (Fig. 6.19f).

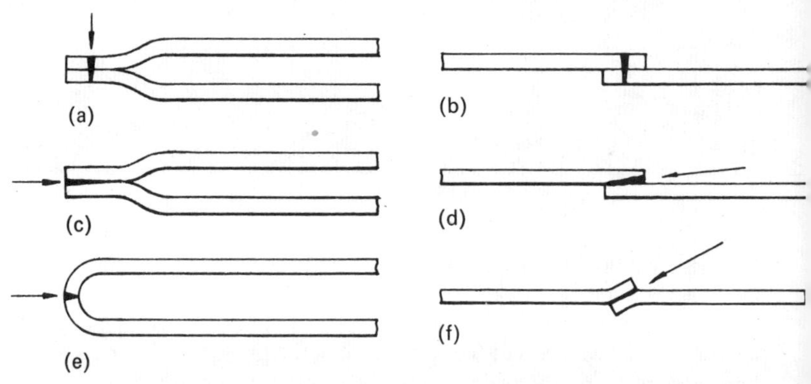

Fig. 6.19 Lap joint design showing progressively better joints (a,c,e), also (b,d,f)

Design for inspection

Whichever joint design is selected there is one overriding factor that has to be considered: the geometry of the lower bead is a function not only of the weld settings but also of the physical properties of the metal near its melting point. Again, lack of complete penetration can be established with certainty only by a visual inspection of the lower bead. It is therefore advisable to design the component so as to make such an inspection possible. This aspect of joint design cannot be over-emphasized for highly stressed applications nor when reliability requirements are stringent. In such cases, if inspection of the lower bead cannot be readily achieved, it may be advisable to consider an alternative joint design.

Economic design considerations

It is appreciated that the capital cost of EBW machines is high by comparison with most conventional welding equipment. The extra cost may be justified for a number of reasons such as the establishment of a superior product or the reduction of rate of scrap. However, when account is taken at the design stage of the special capabilities of the process, direct cost reduction in the manufacture of the product can be secured. The following are examples where, for various reasons, such savings have been obtained.

Sodium-cooled valve

Many components are machined integrally from standard stock, although the design requirements may vary from one part of the component to another. In the absence of a satisfactory joining technique, the highest grade of material demanded by the most stringent requirement will have to be used throughout, with obvious cost penalties. In many instances, savings in material cost can be achieved if the high-grade material is restricted to those areas where it is really required, with cheaper materials used elsewhere.

The valve head shown in section in Fig. 6.20 is subjected to high temperature, the seat to wear, and the upper stem to creep stresses. It was possible to fabricate the valve by EBW so that the head was in a heat-resisting material, the seat in a wear-resisting alloy, and the upper stem in a creep-resisting steel. The conditions to which the lower stem is subjected could be adequately met by the use of a simple low alloy steel.

Bimetallic saw blade

This is a spectacular application which is being commercially exploited and is likely to revolutionize a whole industry. Also, it is an

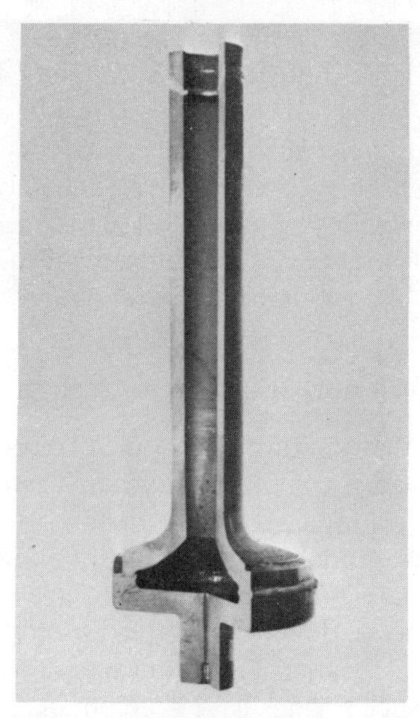

Fig. 6.20 Sodium-cooled valve (*Courtesy Hawker Siddeley Dynamics, Electron Beam Division*)

interesting example because it is not found in an advanced industry, neither are the materials themselves particularly exotic. Nevertheless, the savings are quite substantial and the product technically superior.

A narrow high-speed steel strip is electron-beam welded to the body of the blade which is made from a low alloy steel. The saw teeth are then cut beyond the joining line as can be seen in Fig. 6.21. For a plant with an annual production of 5 000 000ft (1520 000m) of blade, the annual cost saving resulting from the economical use of high speed steel material is £500 000. This obviously far outweighs the capital investment required for a single machine which is capable of handling the total throughput of the plant. The bimetallic saw blade is reported to have better shatter characteristics than one fabricated entirely in high-speed steel.

Cover ring with lugs

Often, components are machined from solid resulting in waste of material, and of operator and machine tool time. It is true that precision forgings and castings can be utilized and that material

Fig. 6.21 Bimetallic saw blade (*Courtesy Hawker Siddeley Dynamics Electron Beam Division*)

waste is then almost eliminated, but such methods are restricted to large quantity production. Because of the precision nature of EBW and the low component distortion, it is often possible to construct certain products by welding their component parts together, the component parts being made from standard stock and with the minimum of machining.

The cover ring with lugs, shown in Fig. 6.22, was fabricated by machining the ring from standard tube stock. The lugs were milled in long straight lengths before being cut to individual short units. This is a simple milling operation as compared with the complexity of machining the whole unit from a tube whose outer diameter embraced the overall dimensions of the component. The lugs were then electron-beam welded to the ring.

Bursting adaptor

Initially, this part was machined from bar and the centre bore drilled to retain a 0·015in. (0·4mm) thick wall at the base of the bore to act as a bursting disc (Fig. 6.23). Considerable difficulty was

Fig. 6.22 Cover ring with lugs (*Courtesy Hawker Siddeley Dynamics, Electron Beam Division*)

experienced in the machining operation and, because of the small tolerance allowed, the scrap rate was unacceptably high.

Now the component is being drilled straight through the bore and discs are punched from solid strip of the exact thickness. These are electron-beam welded in position. The material used is a 13 per cent Cr–2 per cent Ni steel which can meet the stress and corrosion requirements of the component's operation.

Exit duct

This Venturi unit, shown in Fig. 6.24, is part of an aircraft air-conditioning system. The configuration cannot be deep-drawn and spinning was considered the appropriate method of production. However, because of the rapid change in diameter about the middle section of the unit, spinning the unit in one part would have been slow and costly. However, spinning in two sections presented no problem. The two units were then electron-beam welded together, as were the flanges to the Venturi form.

Fig. 6.23 Bursting adaptor (*Courtesy Hawker Siddeley Dynamics, Electron Beam Division*)

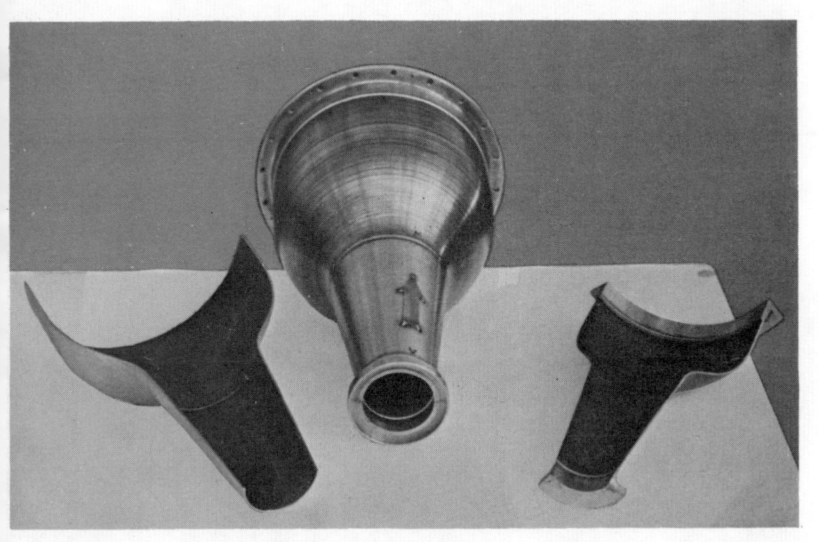

Fig. 6.24 Exit-duct unit (*Courtesy Hawker Siddeley Dynamics, Electron Beam Division*)

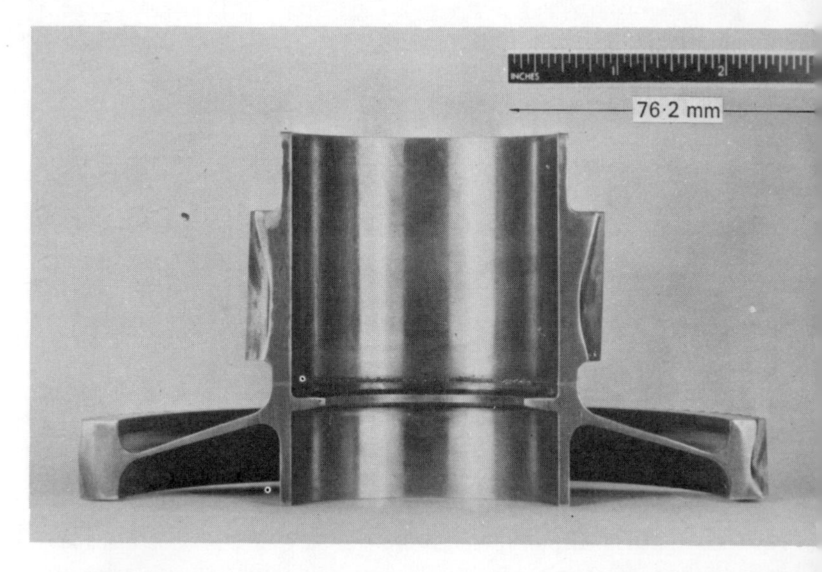

Fig. 6.25 Gear assembly (*Courtesy Hawker Siddeley Dynamics, Electron Beam Division*)

Gear assemblies

Finished gears can be assembled by electron beam without the need for subsequent machining operations. Figure 6.25 illustrates a double helical gear assembled from two gears which had been machined, hardened, and ground separately before welding. This method eliminates the necessity for a wide gap between the two gears; such gaps are unavoidable with integral machining since clearance is required for the hobbing cutter and the grinding wheels used in cutting the gear teeth.

Cluster gears which in the past have been joined together with splines, lock nuts, or bolts can now be welded by electron beam, thus eliminating the fretting associated with mechanical fasteners. There is a significant saving in cost in the application of welded gear assemblies because individual gears can be machined in stacks.

Large crankshaft

When few components are required, forging methods may prove to be uneconomic since the cost of the expensive dies will have to be spread over only a small number of parts.

An interesting example of the application of EBW as an alternative to forging is shown in Fig. 6.26, which illustrates how a crankshaft can be fabricated from piece parts of simple geometry. This concept is particularly suited to large crankshafts such as those used in ships' engines.

Fig. 6.26 Crankshaft assembled by electron-beam welding (*Courtesy Hawker Siddeley Dynamics, Electron Beam Division*)

Fig. 6.27 Relay can (*Courtesy Hawker Siddeley Dynamics, Electron Beam Division*)

Relay can

The relay can illustrated in Fig. 6.27 had been produced by joining the header to the can body by either soldering or arc welding. Several manufacturers have produced this type of component by electron beam over a number of years with greatly improved reliability and significant savings in inspection cost. For example, when soldering was used, each component had to be inspected to ensure hermeticity and freedom from flux contamination, whereas with EBW only occasional spot checks are made.

7. Applications

Difficult-to-weld applications

Electron-beam welding has unquestionably provided a facility to weld materials which by conventional welding techniques are classed as difficult to weld, either because of some unacceptable form of postweld cracking, or severe distortion, or both. It is difficult to separate the metallurgical aspects of welding from the geometrical constraints imposed by the shape of the component; for example, a material may prove to be weldable in an experiment involving a simple butt weld, but cracking may be experienced when welding a component under the constraint of a heavy flange. Again, the inspection standards to which the weld is subjected may themselves be the controlling factor in classifying whether the material is considered weldable, difficult to weld, or even unweldable. Certain parts may be subjected to a mere visual postweld examination but others inspected to a zero-defect standard. Even in the absence of defects, due regard is often given to the geometry of the weld upper and lower beads because of the stress concentration effects of irregularities such as undercutting and notch formation.

Thus, although some materials are inherently difficult to weld, even in the context of essentially simple components, others are classed as difficult. This is not because of basic metallurgical reasons but because of the restraint imposed during welding, either by the component itself or by the fixtures used to ensure acceptable distortion levels in the finished component.

Outlet guide-vane segment

In this application, single vanes in 12–14 per cent Cr steel are to be joined together at the platform; this is to be followed by welding the vane subassembly to a circular plate. Both operations had been previously performed by conventional TIG-welding as the weldability of the material does not present any serious difficulty. However, the reasons for a change to EBW from TIG-welding can be appreciated from a close examination of Fig. 7.1. This reveals the unusual joint configuration where a substantial area of the abutment between the blade platforms could not be welded by TIG techniques owing to the large change in section. Furthermore, the attachment of the fixing flange through a section of 0·150in. (3·75mm) had to be carried out by using a double-chamfering preweld preparation,

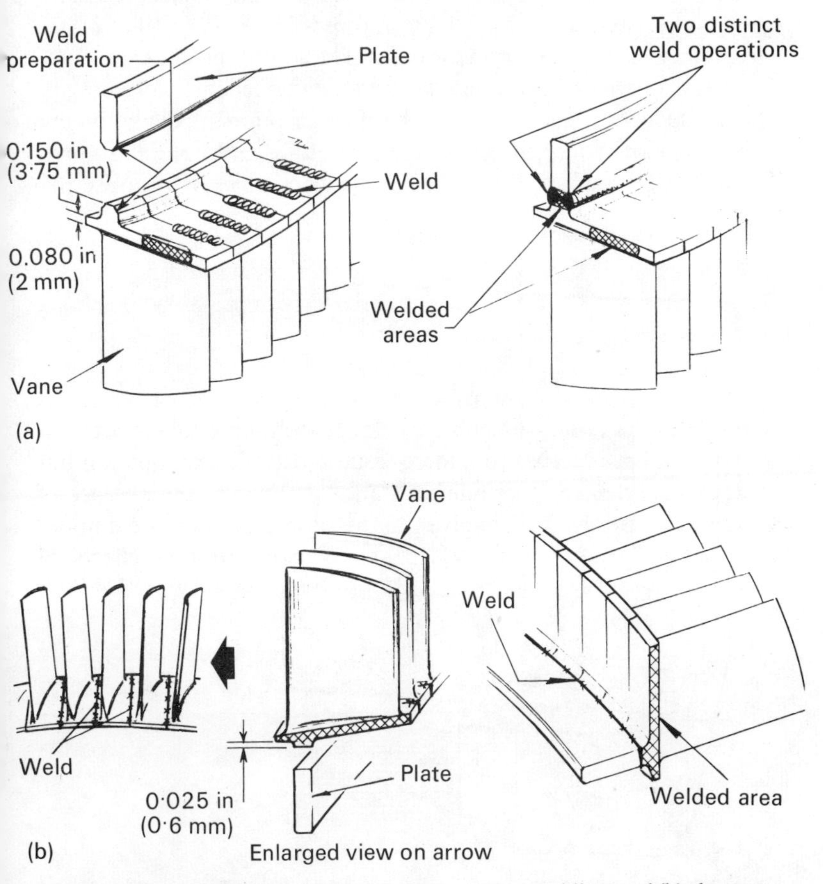

Fig. 7.1 Outline guide vane segment: (a) argon-arc welding, and (b) electron-beam welding (*Courtesy Rolls-Royce Ltd*)

a TIG-weld operation that is not conducive to the important requirement of low distortion.

The application of EBW to this assembly had two immediate advantages. A full-width weld of the platforms could be achieved leaving no areas of the abutment unwelded. Secondly, the thick flange could be attached in one weld pass without preweld chamfering, thus reducing both the cost of machining and the degree of thermal distortion of the welded assembly.

The vanes across the platform faces were welded by beam deflection in preference to a mechanical traverse, since this reduced the function of the manipulator to a simple mechanical indexing movement (see Fig. 7.2). This allowed an examination of the feasibility of this technique for eventual quantity production.

It is probably relevant here to examine the benefits of welding by beam deflection and to give a brief description of the technique used. In a low-voltage EBW system, the gun is mobile and its movement can be combined with that of the worktable. By contrast, in a

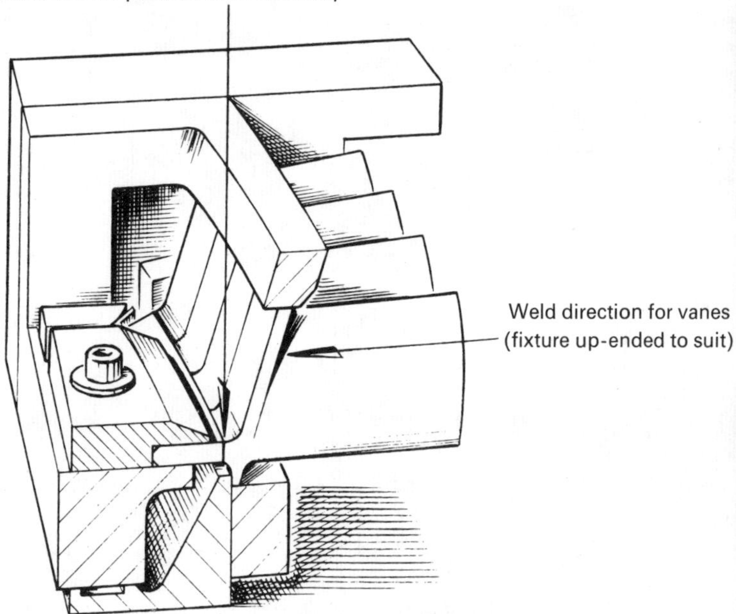

Fig. 7.2 Fixture of electron-beam welding of segment (*Courtesy Rolls-Royce Ltd*)

high-voltage system, the gun is fixed and all manipulation has to be carried out by means of the worktable. Thus, beam deflection, when used for welding with a high-voltage system, affords a limited but often useful beam mobility which compensates for the rigid configuration of a fixed gun. However, for welds of short duration, high welding speeds, or for the welding of profiles with small radii, both systems have similar limitations. Within certain well-defined limits, magnetic beam deflection can have decided advantages. The beam has virtually no mass and so does not present the inertia resistance to movement encountered when manipulating heavily

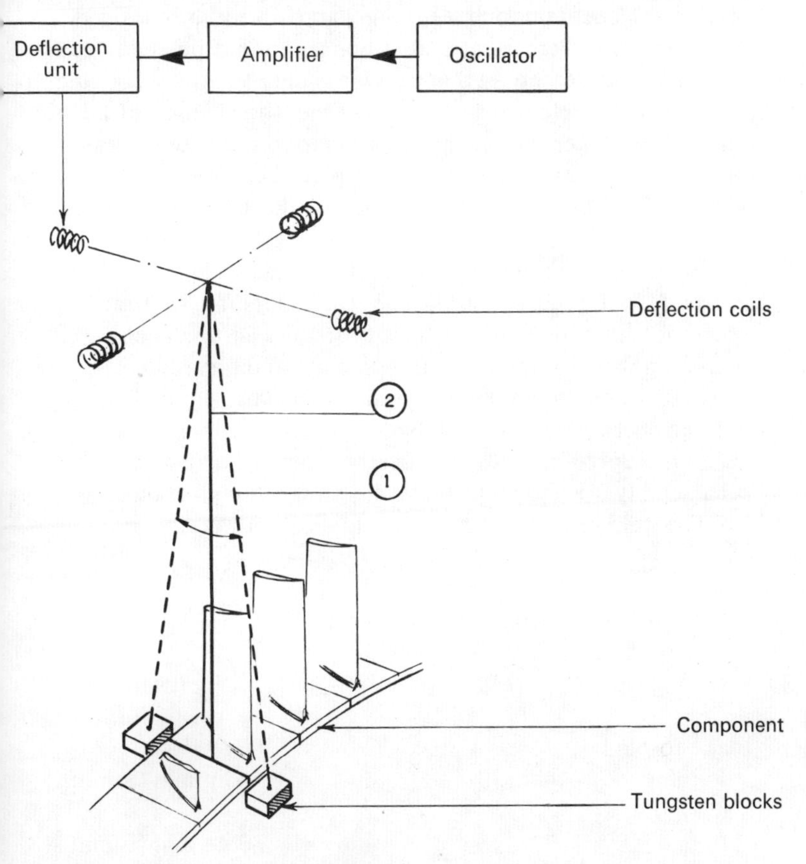

Fig. 7.3 Application of beam deflection: 1—tracing beam: 50Hz low-power beam, locates joint line, and 2—welding beam: fixed oscillation to produce 60in/min (25mm/s) (*Courtesy Rolls-Royce Ltd*)

fixtured components. Again, in a purely mechanical traversing system, it may be difficult to accelerate the part to full welding speed within the limited space in the vacuum chamber. Therefore, although the weld path may be linear, as in the application presented here, beam deflection may offer distinct advantages.

The arrangement for welding by beam deflection is diagrammatically illustrated in Fig. 7.3. Positive indexing on solenoid-controlled stations, although possible, was not advisable, since normal machining tolerances would inevitably result in spacing errors between adjacent weld lines. A technique for accurately placing the beam on each weld line had then to be developed.

The operator brings each weld line to a top dead-centre station using the optical viewing device. This is only a point check, and, to establish that the beam passed along the whole length of the weld line, a low-power tracing beam is made to oscillate at mains frequency between two tungsten blocks placed at the extremities of the weld line. With minor adjustments to the indexing mechanism, the tracing beam is made to travel along the weld line accurately. The weld setting is then selected and the beam is made to deflect at a fixed low-frequency oscillation corresponding to the welding speed of 60in/min (25mm/s). This procedure ensures that a weld of minimum width is applied with reliability and precision.

Attaching the fixing flange to the vane subassembly is a simple welding operation carried out by rotating the component so that the weld line is presented to a stationary beam. Two views of the vane subassembly are shown in Fig. 7.4, and the complete block in Fig. 7.5. The weld settings used for this application were:

Operation	*Accelerating voltage (kV)*	*Beam current (mA)*	*Welding speed*
Blade-to-blade, tracing	130	0·5	50Hz (beam oscillation)
Blade-to-blade, welding	130	10	60in/min (25mm/s)
Vane assembly to flange	135	8	25in/min (10mm/s)

This application illustrates a number of useful features of welding by electron beam. A satisfactory weld between the blades was made

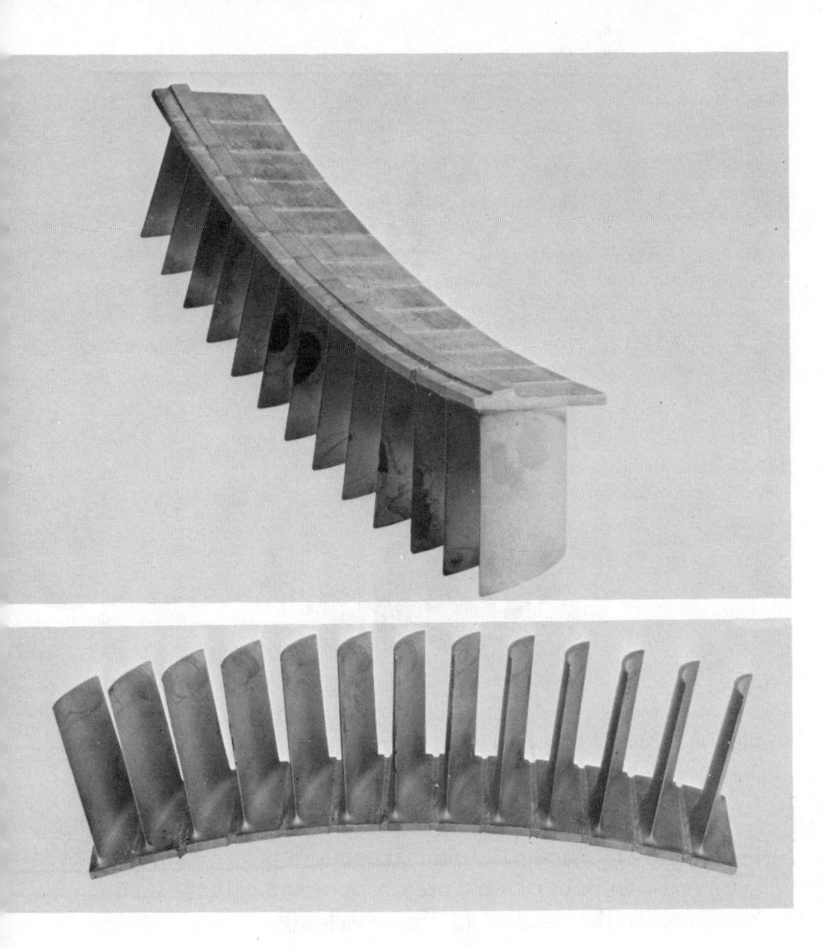

Fig. 7.4 Welded vane (*Courtesy Rolls-Royce Ltd*)

possible by the ability of the beam to produce welds in areas of poor accessibility. In TIG-welding, the weld line between the flange and the vane platforms had to be 0·150in. (3·75mm) below the platform face. In EBW this was reduced to 0·025in. (0·6mm) and it was possible to devise a weld setting which gave full penetration for the whole width of the abutment, including the locally thickened section. The integrity of the component was therefore considerably improved since no part of the abutment was unwelded and there was no potential source of failure.

Fig. 7.5 Complete segment (*Courtesy Rolls-Royce Ltd*)

The deep penetration capabilities of EBW were utilized in welding the flange. The 0·150in. (3·75mm) penetration was readily achieved in one pass without preweld preparations and with the minimum of heat input to the component during welding.

Hollow sheet-metal vane in titanium alloy

Many reasons for adopting EBW have been discussed: welding of metallurgically 'difficult' alloys, intricate joint configurations, and so on. The application described in this section indicates an additional justification: a change of material in a component which had hitherto been successfully manufactured may render the method of manufacture unsuitable.

Sheet-metal aerofoil vanes in 12–14 per cent Cr steel and 18/8 stainless steel have been successfully produced in large numbers. A combination of resistance welding of the trailing edge and high-temperature brazing of an internal supporting tube is generally adopted. The introduction of a titanium alloy, IMI230 containing 2 per cent Cu, made brazing an irrelevant process and an alternative design and method of manufacture had to be evolved.

A two-piece construction was selected, shown diagrammatically in Fig. 7.6. The two halves were machined from thick stock to produce two abutting surfaces where the support tube would normally be placed: these are electron-beam welded, as well as the leading edge of the vane. The trailing edge is first closed in by electron beam and then seam welded.

In the two weld stations inside the vane shown in Fig. 7.6, both abutment faces are some 0·08in. (2mm) deep and 8in. (200mm) long. The illustration also indicates the ability of the electron beam to operate in areas of extremely limited access. The abutment nearer the leading edge is at a depth of 0·500in. (12·7mm) and the beam passes through a 0·060in. (1·5mm) gap. The weld near the trailing

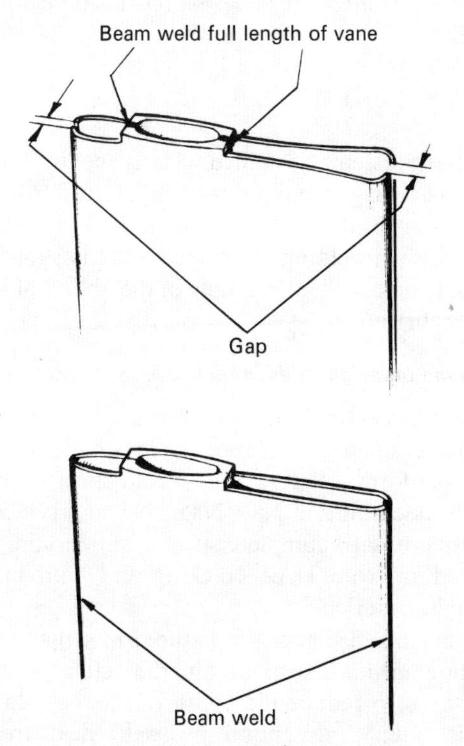

Fig. 7.6 Design of fabricated vane (*Courtesy Rolls-Royce Ltd*)

edge was carried out at a depth of 1·500in. (37·5mm) and through a gap of 0·100in. (2·5mm). A high-voltage well-focused beam is preferred for such an application; in fact, a Zeiss 150kV, 3kW machine was successfully used for this work.

The two welds contained within the aerofoil were accurately machined to fall in a common path, and could have been, at least theoretically, simultaneously welded. Indeed, in a less complex material this would have been feasible. For consistent reproduction of geometrically acceptable weld configurations, however, two separate welds were undertaken. Weld surface and underbead geometry when welding thin sections in titanium alloy are particularly sensitive to welding parameters.

The weld settings used were:

Operation	*Accelerating voltage (kV)*	*Beam current (mA)*	*Welding speed (in/min)*	*(mm/s)*
Internal welds	100	4	24	10
Leading and trailing edges	100	2	24	10

The resultant assembly, which clearly could not have been welded by any existing technique, was singularly free from distortion and the weld survived the forming operations to which the vane was subjected after welding. A formed vane is shown in Fig. 7.7, which is clearly an excellent example of the ability of EBW to open new areas of fabrication.

Welding turbine blade shrouds in nickel-base alloys

A requirement has existed for many years for assembling groups of turbine blades in the higher Nimonic alloys by fusion butt-welding the shroud platforms. Before the introduction of Nimonic 115, turbine-blade assemblies in N80, N85, N95, and N105 were welded by argon arc, not with complete success but certainly with an acceptable freedom from HAZ cracking and with many years of successful flight experience.

The fully machined blades are fixtured in a manner compatible with the engine turbine disc fixing, and the weld is produced without filler wires so as to reduce the width of the weld to a minimum. Wider welds often crack upon postweld heat treatment. The sectional thickness at the weld line is normally in the range 0·040–

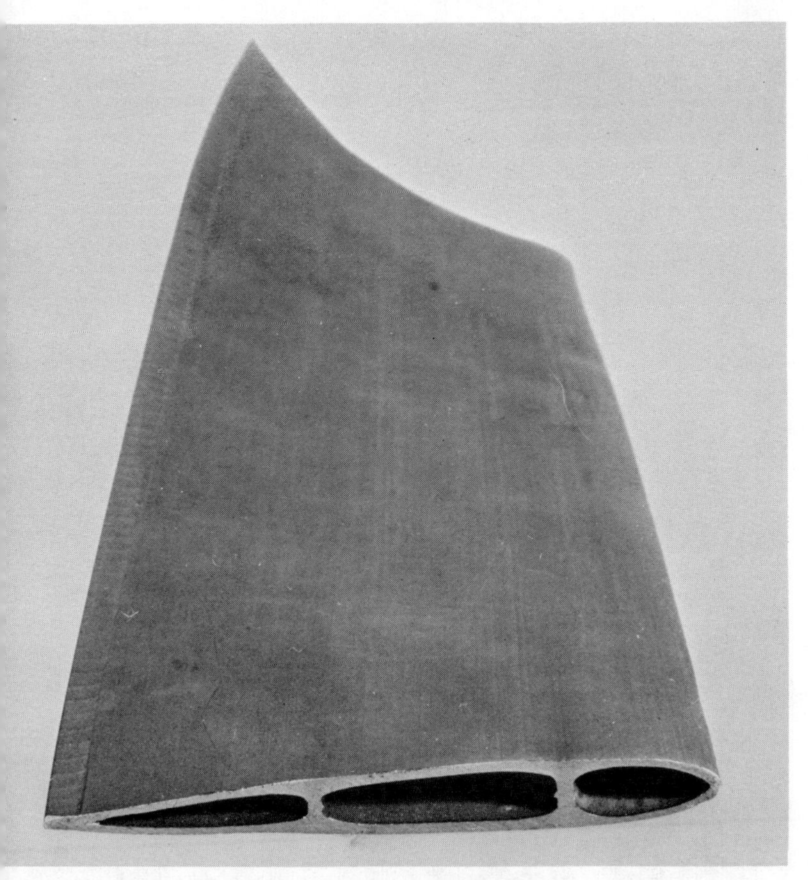

Fig. 7.7 Welded vane (*Courtesy Rolls-Royce Ltd*)

0·050in. (1–1·25mm). The presence of gas seals integral with the blade shroud presents a problem in that their local removal to allow a clear path for the argon arc torch naturally results in a reduction in the efficiency of gas sealing. Alternatively, welds of limited length carried out between the seals would inevitably fail owing to the restraint imposed at the unwelded sections because of the shrinkage stresses induced by the TIG welds.

With the introduction of Nimonic 115, there was an immediate awareness that even the best mechanized procedures devised for TIG-welding would be incapable of producing crack-free welds. Electron-beam welding was the obvious alternative.

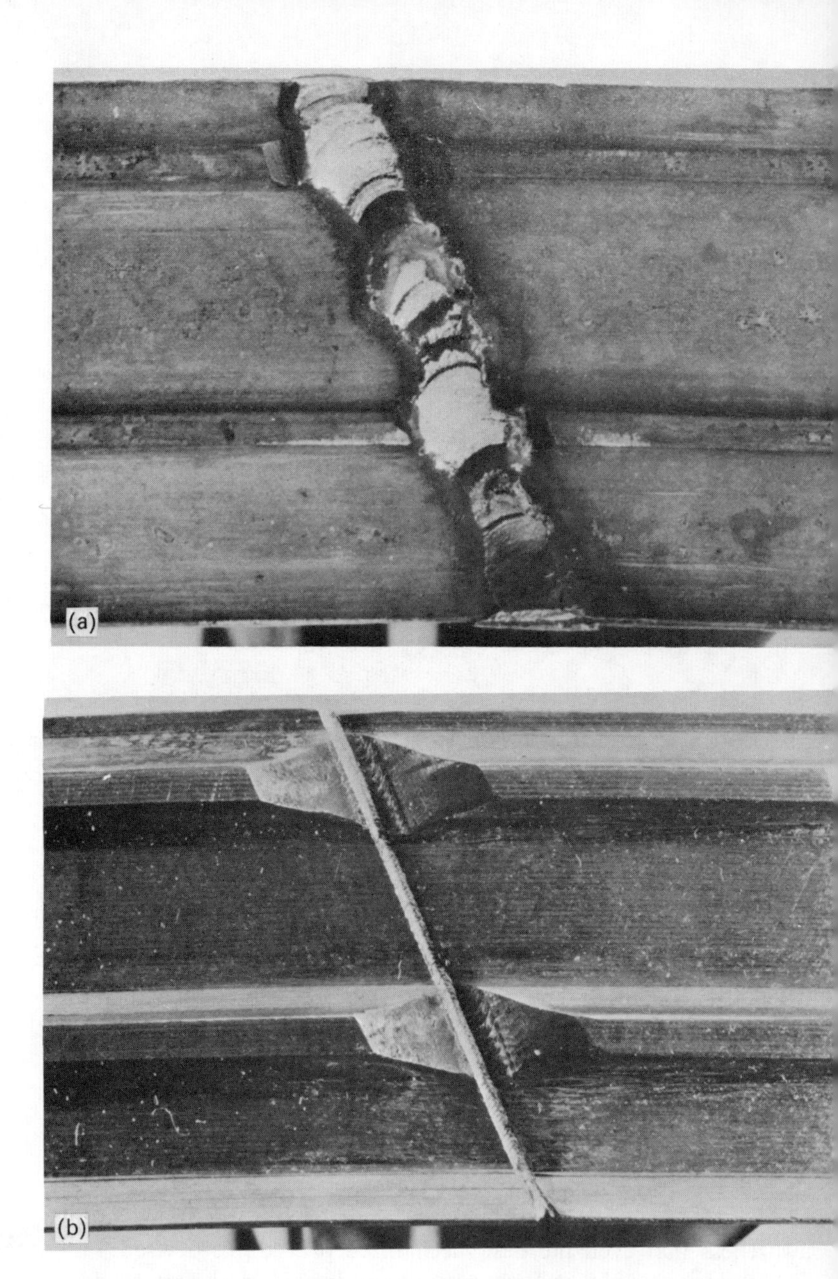

Fig. 7.8 Blade shrouds showing comparison between (a) arc, and (b) electron-beam welding (*Courtesy Rolls-Royce Ltd*)

It was appreciated that EBW had two distinct advantages. Firstly, the very narrow welds would result in a significant reduction in transverse shrinkage stresses, and, secondly, the length of sealing to be removed to facilitate access for the electron beam would be substantially shorter than that required for TIG-welding. The comparison shown in Fig. 7.8 is self-evident. It must be pointed out, however, that the TIG weld was produced manually with an added filler to improve the ductility in the fusion zone. The electron-beam weld, although undertaken without filler, was completely free from cracks. The length of seal removed in both cases in Fig. 7.8 was in accordance with the standard demanded for TIG-welding and is certainly more than necessary for EBW. Narrower total gaps of some 0·025in. (0·6mm) were found to be adequate in the latter case, as can be seen in Fig. 7.9.

Fig. 7.9 Narrow weld with minimal seal interruption (*Courtesy Rolls-Royce Ltd*)

The welding fixture used in this application is shown in Fig. 7.10. The blades are held in a manner simulating the engine arrangement in which they are attached to the turbine disc utilizing a fir-tree configuration. The operator views two assemblies in the same fixture with the two weld lines on a common axis. The cavity in the centre retains a small piece of tungsten on which the operator establishes a sharp beam focus before welding.

To sum up, EBW presented many advantages in this application: better gas-sealing efficiency results from the minimal interruption

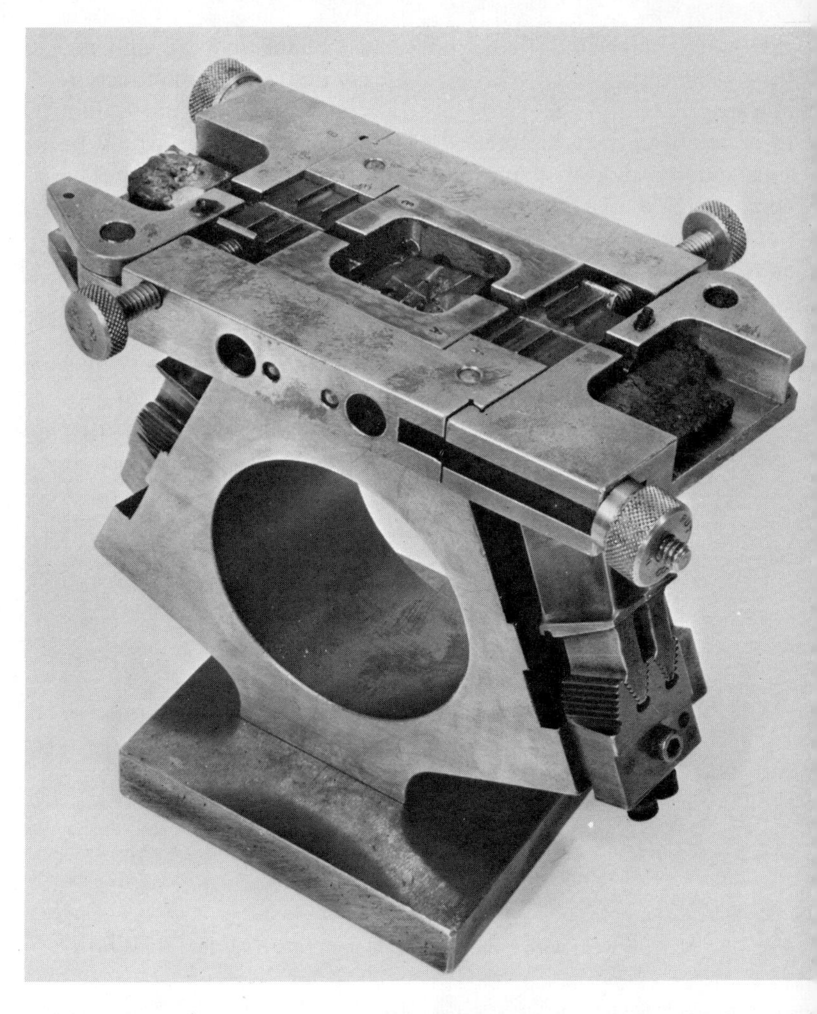

Fig. 7.10 Fixture for welding two pairs of blade shrouds (*Courtesy Rolls-Royce Ltd*)

to the sealing members; the fir-tree root type of fixing the blades to the disc demands an exceptionally high degree of positional accuracy, and the freedom from distortion brought about by the application of EBW is a major advantage. Finally, Nimonic 115 must be considered unweldable by any known fusion welding method other than EBW.

Alternative method avoiding seal removal

Although the reduction of the gap in the sealing fins is an advantage with EBW, there must be occasions when any break in the seal member would be considered objectionable. In these instances EBW can reveal one of its major advantages, that of welding through thick sections with no preweld preparation nor the addition of filler wire.

The welding operation can be carried out at the stage when the shroud is in its rough-machined condition (Fig. 7.11a). The thickness at the weld line is some 0·2in. (5mm) and a gap-free abutment is more readily achieved than in the fully machined shroud condition. The combined shrouds are then machined to give a completely intact seal, as can be seen in Fig. 7.11b.

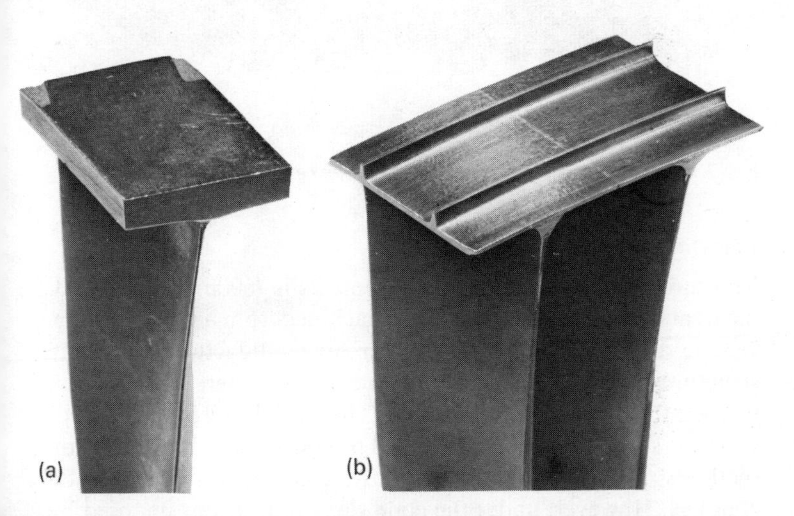

Fig. 7.11 (a) Shroud before machining, and (b) pair of welded shrouds with no seal interruption (*Courtesy Rolls-Royce Ltd*)

The welding of Nimonic 115 in sections greater than 0·250in. (6·25mm) involves certain difficulties at present: fine microcracks are sometimes evident in the HAZ. Future developments will include preheating techniques, the addition of filler to increase the local ductility of the alloy, and an exploration of the effects of a wide range of welding speeds.

Fig. 7.12 Turbine casing (*Courtesy Rolls-Royce Ltd*)

Planetary welding of turbine casing

The conical casing in Fig. 7.12 is some 24in. (600mm) average diameter and 8in. (200mm) deep. It is machined from a single forging in Jethete, a Cr steel used extensively in the manufacture of fabricated structures in modern gas-turbine engines, where the operating temperature does not exceed 450°C. The weldability of this and other Cr steels such as Rex 607 is excellent by conventional fusion-welding methods, provided the geometry of the component is not unduly complex. However, under the conditions of restraint imposed by a complex structure, or a rigid fixture intended to reduce thermal distortion, cracking may occur.

A cross-sectional view of the casing is shown in Fig. 7.13 in which the ringed area indicates a retaining lip machined integral with the casing. During service, the lip sustains damage from 'burning' in the hot gas stream. The salvage procedure selected was to machine the lip to the middle of the narrow section as can be seen in the enlarged view. A replacement ring in the same material then had to be welded into position. This required a planetary weld which

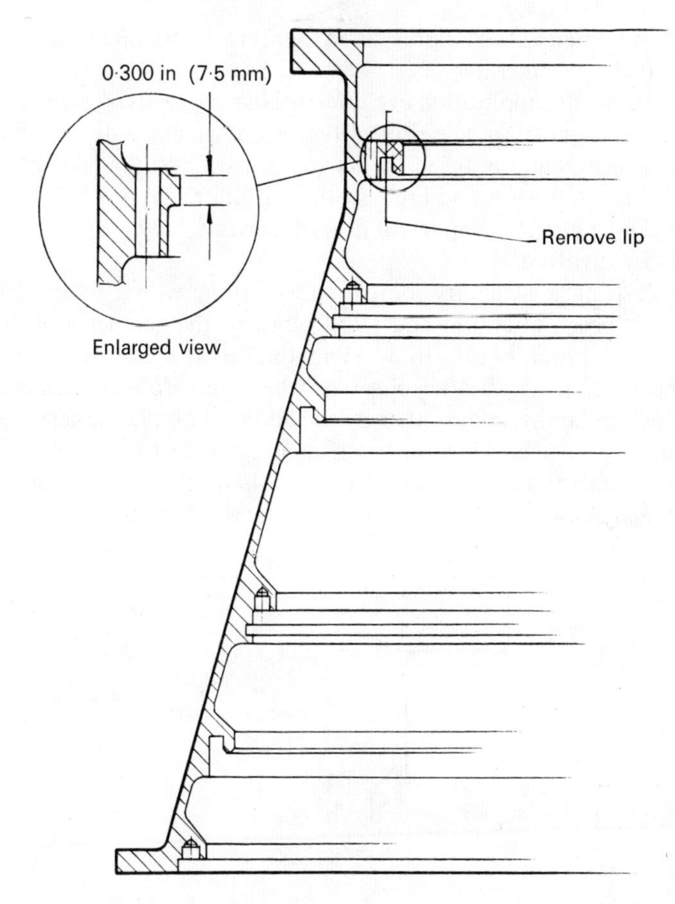

Fig. 7.13 Casing before removing damaged lip (*Courtesy Rolls-Royce Ltd*)

is considered to be one of the more difficult methods of fusion-welding.

Maintaining a gap-free abutment over the full circumference is really the essential condition for success in producing a planetary weld. Massive fixtures and tack welding are often used to ensure against the generation of an unacceptable gap towards the final stages of the weld periphery. Gaps are naturally produced by the shrinkage of the weld metal upon solidification. Conventional techniques producing comparatively wide welds result in large internal stresses leading to cracking during the postweld heat treatment. Even at the welding stage itself, cracks are often detected in the end-of-weld crater.

The successful application of EBW to planetary welds is essentially due to the high energy density of the beam, which results in high welding speeds and reduced shrinkage forces. Thus, the characteristically narrow welds and high depth-to-width ratios of EBW have opened up a new and important field of fabrication where planetary welds are involved.

In devising a planetary joint for the turbine casing under discussion, special attention had to be given to the geometry of the upper and lower beads. In welding thicker sections, as in this example, a concavity is often observed when using EBW (see chapter 6). This can be overcome either by the addition of filler material or by the use of a backing plate which is machined after welding. The latter method was adopted, and Fig. 7.14 illustrates the arrangement used. A fairly substantial ring in Jethete was designed to provide

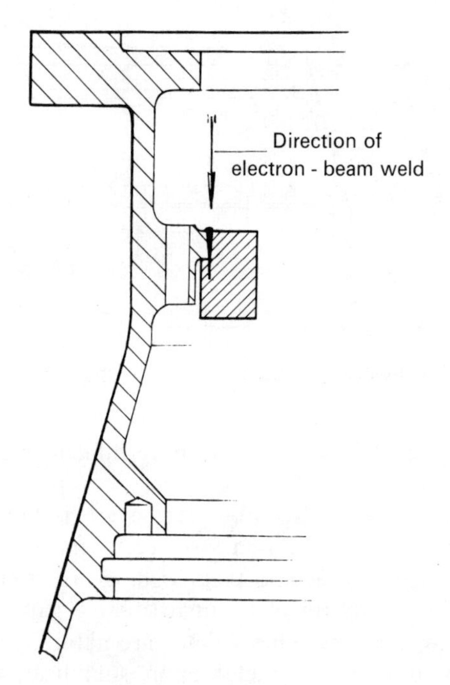

Fig. 7.14 Electron-beam welding of replacement lip (*Courtesy Rolls-Royce Ltd*)

both the replacement lip and a backing plate. To ensure the all-important preweld fit-up, the ring was machined to a 0·004–0·008in. (0·1–0·2mm) interference fit on diameter. For assembly, the casing was heated to 150°C. While the assembly was still hot, a simple boiler plate bung was fitted into the inner diameter of the replacement ring; the interference fit in this case being from zero to 0·004in. (0·101mm).

To ensure that any porosity in the root of the weld bead was completely removed by subsequent machining, the depth of strike of the beam was controlled to be some 50 per cent greater than the final thickness of 0·300in. (7·5mm). A suitable setting was devised using linear testpieces of identical joint configuration. The weld parameters used were:

Accelerating voltage (kV)	*Beam current (mA)*	*Welding speed (in/min)*	*(mm/s)*
140	16	50	20

The end of the weld was allowed to overlap the start of the weld by 0·125in. (3·1mm). A beam current decay device was introduced to perform as a crater filler, thus eliminating end-of-weld cavity cracking.

The welded assembly received a postweld tempering treatment at 570°C for 1h, followed by final machining to restore the original lip dimensions as shown in Fig. 7.15. Dimensionally, the casing was found to be entirely free from distortion and the integrity of the weld was demonstrated by a satisfactory radiographic and dye-penetrant examination.

Turbine casing with nickel-base replacement ring

The successful operation described above suggested that the life of the component could be significantly increased if the replacement ring could be made from a more heat-resisting material, say a nickel-based alloy. Predictably, a material combination in the weld of a nickel-based alloy and a Cr steel would produce a weld nugget of similar properties to 18/8 stainless steel which has lower strength. Following a series of tests, Nimocast 242 was selected for the replacement ring. This alloy had been developed for investment casting applications and has good resistance to both creep and thermal shock in the temperature range of 1000–1050°C.

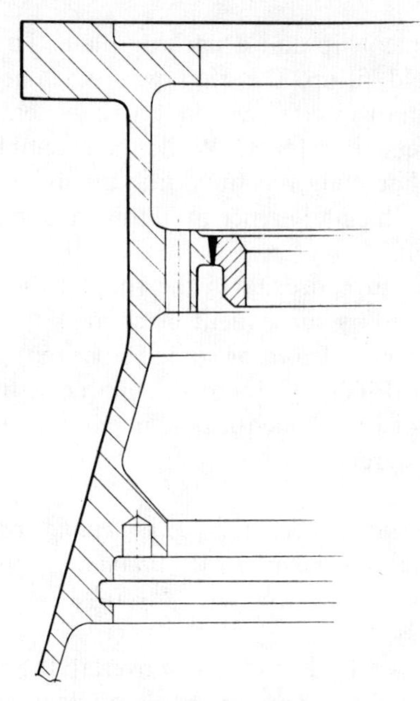

Fig. 7.15 Lip machined after welding (*Courtesy Rolls-Royce Ltd*)

Since the physical properties of Nimocast are different from Jethete, the weld parameters had to be modified from those given above. In every other respect, however, the same procedure as described above was followed. The new welding parameters were:

Accelerating voltage (kV)	*Beam current (mA)*	*Welding speed (in/min)*	*Welding speed (mm/s)*	*ac beam oscillation*
140	16	25	10	0·015in. (0·4mm) amplitude

The transverse beam oscillation was superimposed on the weld setting so that an improved weld appearance might be achieved. A substantial increase in the life of the component is now anticipated.

Clearly, this application could not have been undertaken by conventional welding methods. A successful salvage operation of an expensive component was achieved, and a new design concept was evolved, both resulting in a superior component.

Deep penetration planetary weld in creep-resisting ferritic stainless steel

A simulated turbine disc in FV535 was subjected to a rig-testing schedule which had generated cracks from the holes near the outer periphery. Arrows indicate two cracks on the disc shown in Fig. 7.16. It was considered well worth while to undertake a salvage operation involving the removal of the defective outer rim and fusion-welding a new rim to the existing disc. Because of the severe dimensional requirements demanded from the as-welded condition of the piece and the inherent difficulty of using conventional welding methods to meet the metallurgical requirements of FV535, electron-beam welding was considered to have a better chance of success. A satisfactory welding technique was then evolved.

Fig. 7.16 Simulated turbine disc; arrows indicate test cracks (*Courtesy Rolls-Royce Ltd*)

Reference has previously been made to the inherent difficulty of producing a fusion weld in a planetary weld configuration, mainly owing to the difficulty in preserving an acceptable gap-free abutment throughout the duration of the peripheral weld. The sectional thickness at the weld line was 0·300in, (7·5mm) and the work had to be carried out on a Zeiss machine of 3kW capacity. If full penetration had to be achieved in one pass with the limited power available, a relatively low welding speed, in the region of 10in/min (4mm/s) would have had to be selected. This would have probably led to the generation of an unweldable gap in the joint line at the final stages of the weld and to an unacceptable degree of distortion of the disc. It was appreciated that a higher power gun, of either a low or high voltage, would have given complete penetration in one pass at high welding speed. However, with only 3kW available, a technique involving welding from both sides was developed, as it was thought to have a better chance of producing a disc with the least distortion.

An additional useful feature of welding from both sides in a fairly thick section is that such a procedure would improve the geometry of the top weld beads. Since the original thickness of the disc could not be reduced, it was essential to produce a weld bead free from concavity and undercutting. A weld of partial penetration is analogous to a weld penetrating through to a backing plate. Thus, in a weld produced by two passes, one from each side, the backing support afforded to the penetrating weld bead ensures a positive 'crown' on both sides. Attention must be paid, however, to the possible occurrence of root porosity, although it would appear that if any porosity exists from the first weld, it is often 'filled in' by the passage of the molten metal during the second weld.

However, there is an inherent disadvantage in welding a thick section from both sides. This is related to the manner in which the two welds meet about the midway point and to the establishment of an acceptable inspection technique which would reveal any lack of fusion if the two welds did not meet (Fig. 7.17a). In Fig. 7.17b, although the two welds meet, the narrowness of the 'dagger-like' welds may lead to a portion of the abutment being left unwelded. Very narrow welds, although ideal from the point of view of reducing thermal distortion, are difficult to apply with precision, particularly with existing systems which rely solely on operator skill in aligning the beam with the weld line.

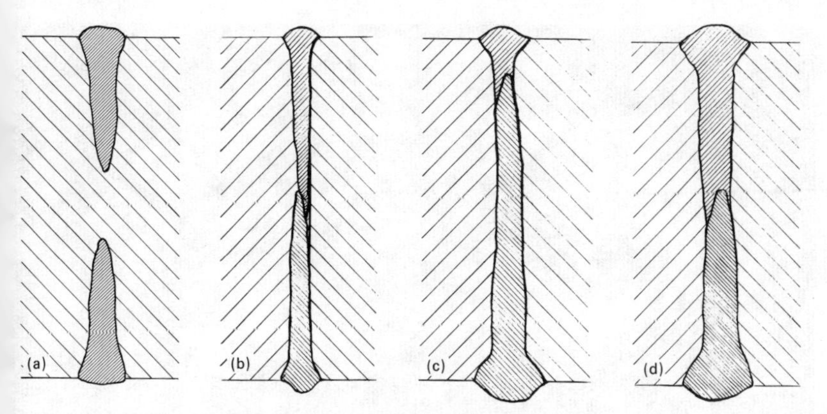

Fig. 7.17 Alternatives of double weld (*Courtesy Rolls-Royce Ltd*)

Although the condition shown in Fig. 7.17c is free from internal defects, it is probably unacceptable because of the uneven distortion which may result from the fact that more heat is applied from one side. A condition as illustrated in Fig. 7.17d is the most acceptable since there is complete fusion of the abutment and the welds are wide enough to ensure against any minor alignment errors. The settings used for both operations were:

Accelerating voltage (kV)	*Beam current (mA)*	*Welding speed (in/min)*	*(mm/s)*
135	15	25	10

One further virtue of adopting the technique of welding from both sides results from the restraint produced by the first weld. Had the full penetrating beam power been applied, it might not have been possible to prevent an unweldable gap forming towards the end of the weld run. In effect, the first weld acted, in addition to its main structural task, as a tack weld for the second run.

The salvaged finished disc is shown in Fig. 7.18; it was remarkably free from distortion and has been impressively successful in extended postweld rig-testing schedules.

An unusual feature of this salvage was that the disc could not be subjected to any heat treatment after welding. This would have relieved the stresses accumulated in the centre of the disc, thus nullifying the object of the cumulative rig-test schedule. Tempering would have been an obvious advantage from the welding point of

Fig. 7.18 Simulated disc after welding (*Courtesy Rolls-Royce Ltd*)

view as it would have increased the ductility of the weld. The test on the disc had to be continued in the as-welded condition without tempering, but this was obviously a special case.

Electron-beam welding has been treated in the aero-engine industry precisely as any other fusion welding process; so, where postwelding heat treatment has been applied to TIG-welded assemblies, electron-beam welds have been identically processed. Whether or not a special case can be made out for an easement of heat treatment requirements for electron-beam welds for certain materials is still an open area of research, not as yet fully explored.

The only safe procedure that can be recommended for the present is to follow the same rules as laid down for conventional fusion-welding techniques.

It may be of interest here to record the relative mechanical properties of FV535 in the two conditions encountered in the turbine disc.

Condition	*UTS* (*tonf/in²*)	(*MN/m²*)	*Elongation* (*per cent*)
Parent material	69	(1070)	20
As-welded	69	(1070)	5
As-welded and tempered	73	(1130)	16

Welded compressor rotor shafts

Once the capabilities of EBW become generally known by the designer, he may be tempted to introduce certain joint configurations, which, although presenting no problems with regard to welding, may be difficult to inspect for weld quality. Such a temptation is more likely to occur when the designer is dealing with a complex component requiring many separate welds. A study of some of the welding and inspection problems involved in assembling a light-alloy rotor will indicate that, even with the special capabilities of EBW, considerable thought has to be given to its correct application.

A great many welded rotor shafts have been satisfactorily welded by argon arc over a number of years. A change to EBW was justified only on the grounds of an improved postweld dimensional control, freedom from a high incidence of porosity which was a problem with TIG-welding, and an improved repeatability resulting from a reduced reliance on operator skill. It will be clear from the last of these considerations that, in the absence of a weld-line-sensing device, the operator has the ultimate responsibility for ensuring that the beam is placed precisely on the abutment. The weld settings should be carefully developed so as not to produce too narrow a weld; the narrower the weld, the more critical the beam-positioning operation becomes.

The light alloy used in the assembly under consideration is a copper-bearing heat-treatable alloy, Hiduminium 54. The separate discs were fully heat treated before welding and the completed assembly aged only after welding.

The construction of a typical rotor is shown in Fig. 7.19, which also indicates the type of location of the separate discs which make up the assembly. Some form of location is required to ensure concentricity of the assembly after welding, since the individual discs are fully machined before welding and no postweld machining can be carried out. The postweld dimensional requirements are a total eccentricity reading of no more than 0·005in. (0·125mm) and a 'swash' in any of the blade retaining rims not exceeding 0·008in. (0·2mm).

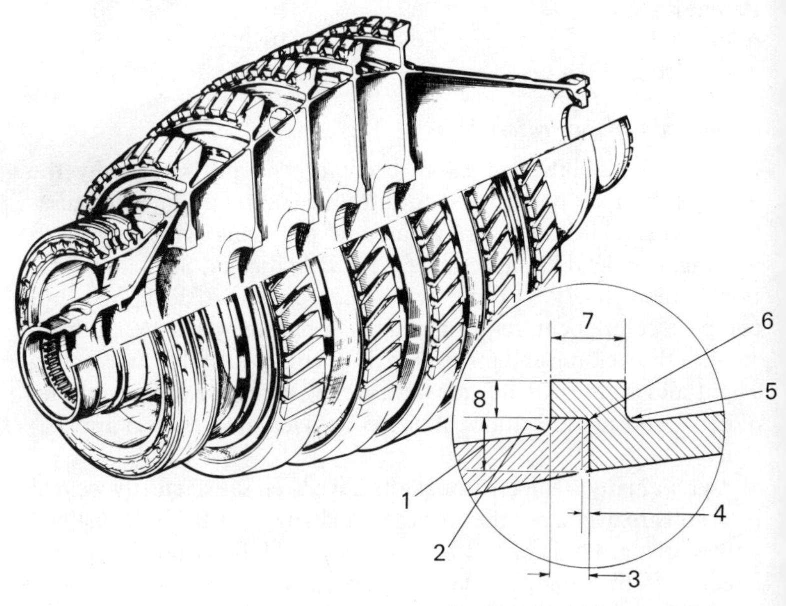

Fig. 7.19 Rotor shaft assembly: 1—0·067in. (1·67mm); 2—0·015in. (0·38mm) rad. max; 3—0·050in. (1·25mm) (−0·005[0·13]); 4—0·005in. (0·13mm) × 45° (+0·005[0·13]) ± 2°; 5—0·015in. (0·38mm) rad. max; 6—0·005in. (0·13mm) rad. max; 7—0·100in. (2·54mm) (−0·005[0·13]); 8—0·050in. (1·25mm) (−0·005[0·13]) (*Courtesy Rolls-Royce Ltd*)

Meeting the concentricity requirement presented no problems with TIG-welding. However, in the region of the end of the weld, local swash is induced because of the over-run with the start of the weld and because of crater-filling requirements. Naturally, the characteristically wide argon-arc weld contributes to the problem and this was an area in which EBW could well be expected to produce more acceptable results.

The location device shown in Fig. 7.19, which has been successfully used for many years in association with TIG-welding, performs two functions. Primarily, it is a fixturing arrangement, but it also provides the means of adding a controlled volume of filler material to the weld. However, during EBW, the weld line is naturally obscured. This difficulty can be surmounted if a mark is introduced on the outer perimeter of the filler 'lip'. The mark can be a continuous line, although an accurately positioned centre-punch mark may suffice. The greater hazard in obscuring the weld line is the inability to observe any opening up of the abutment because of progressive shrinkage distortion. Such opening up naturally adds to the run-out problems at the end of the weld. It is essential therefore to preserve a 'hard end-on' abutment throughout the weld cycle.

The spigot fit must be based on some form of interference, since a clearance fit, however small, would induce eccentricity problems. However, an excessive interference would prevent the essential gap-free abutment at the joint line from being maintained, particularly when welding materials such as an aluminium alloy where both the thermal conductivity and coefficient of expansion are high. When the preweld machining difficulties and the welding requirements outlined above are taken into account, a fit not exceeding 0·002in. (0·05mm) inteference was adopted.

The following are typical weld settings for a penetration of 0·080in. (2mm):

Accelerating voltage (kV)	*Beam current (mA)*	*Welding speed*
120	13	120in/min (50mm/s)

The welding speeds selected for this application are obviously high even by EBW standards. It was established that such speeds were particularly beneficial in reducing the incidence of porosity in aluminium alloys.

A general view of the assembly being prepared for welding is shown in Fig. 7.20 and a finished component in Fig. 7.21. Although the working distance for the six welds involved varied between 9 and 14in. (225–350mm), no alteration to the weld settings was necessary. Naturally, however, the beam had to be refocused at the height of each weld line.

Fig. 7.20 Rotor shaft being loaded into vacuum chamber (*Courtesy Rolls-Royce Ltd*)

POSSIBLE ALTERNATIVE CONSTRUCTION. It must be appreciated that the combined locating filler arrangement described above is both expensive and difficult to produce to the required tolerances. Some less stringent methods of location were considered with EBW, particularly since thermal distortion and shrinkage displacement are both of a substantially smaller order than in TIG-welding.

Perhaps the simplest form of location is an internal spigot which, from the point of view of EBW, has one or two advantages. The weld abutment is not obscured and the arrangement provides support to the weld, thus assisting in forming an upper bead free of concavity or undercutting. However, there are certain disadvantages, particularly for components demanding high integrity. Firstly, the non-welded portion of the spigot may act as a stress raiser. Also, the presence of the spigot impedes a complete inspection of the lower bead. However, if the geometry of the component allows the machining of the spigot after welding, the objections just listed

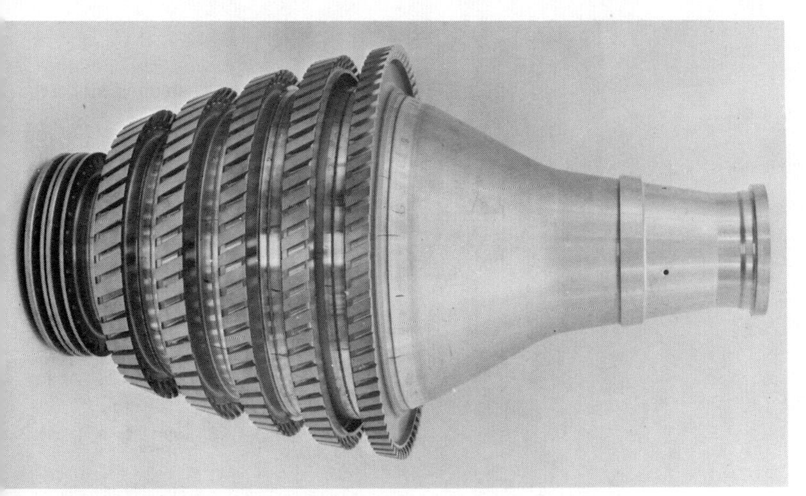

Fig. 7.21 Welded rotor shaft (*Courtesy Rolls-Royce Ltd*)

are then of no consequence. In the case under discussion, the geometry of the component made it virtually impossible to machine the spigot, and other alternatives had to be considered.

A possible arrangement is illustrated in Fig. 7.22 utilizing a square abutment without recourse to internal or external location. The weld section was locally thickened from 0·080–0·100in. (2–2·5mm) so as to permit the acceptance of some small degree of top bead concavity, or alternatively to allow the machining of such irregularity while still retaining the full nominal section of the rotor. The separate discs were located by means of a solid mandrel fitting the centre bores of each disc. Loading was achieved by the action of strong coil springs which ensured an acceptable gap-free abutment at each weld and throughout the complete welding cycle, despite the tendency for gaps to occur during welding owing to shrinkage displacement, which, for Hiduminium 54 at a weld thickness of 0·100in. (2·5mm), was found to be 0·004in. (0·1mm) for each weld.

A rotor welded under the above conditions is shown in section in Fig. 7.23; there is a well-formed underbead and a surface concavity of less than 0·005in. (0·13mm) which could be readily corrected by machining.

The expected improvement of using EBW did, in fact, materialize in the above application. The principles outlined above can therefore be applied to other materials and to thicker sections. Due attention

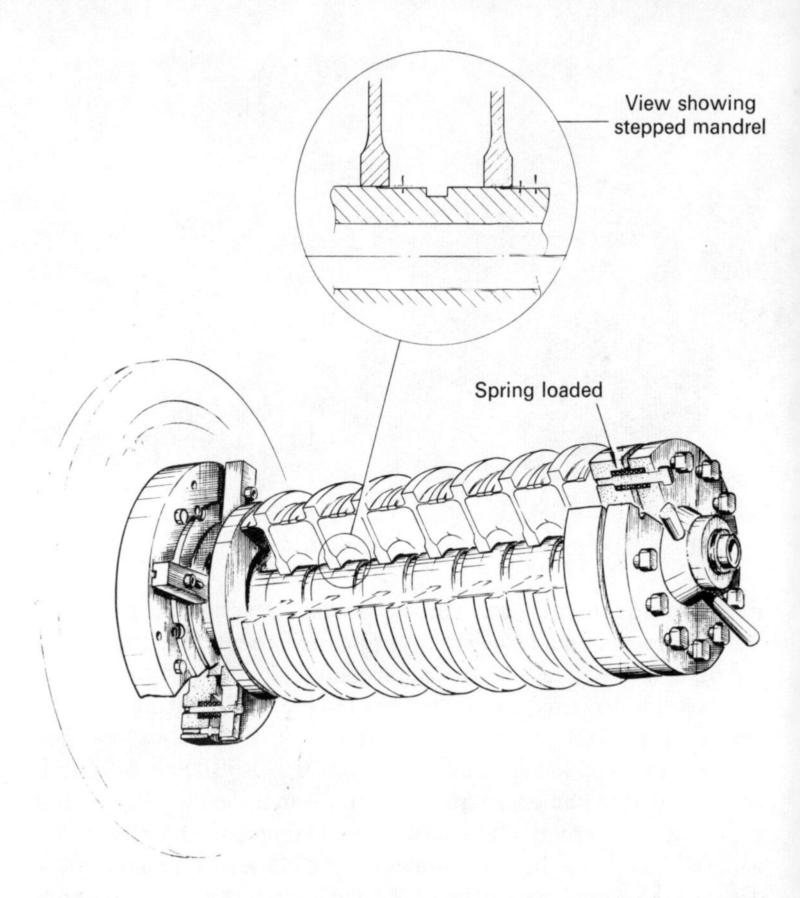

Fig. 7.22 Fixture for assembling multidisc rotor (*Courtesy Rolls-Royce Ltd*)

must be paid, however, to the development of suitable joint configurations and to inspection techniques.

Welding small components

The advent of electron-beam welding has been particularly useful in the field of instrumentation where small welds on precision components are often replacing the old-established methods of mechanical fastening, soldering, or the use of adhesives. When applied intelligently, the technique confers advantages of cost and reliability.

Fig. 7.23 Welded rotor, cross-sectional view (*Courtesy Rolls-Royce Ltd*)

Although the general advantages of EBW have been enumerated in other chapters, the following features are particularly relevant to the manufacture of instruments and other small devices.

The local fusion resulting from high beam power density produces small weld nuggets with confined HAZs. It is thus possible to weld close to temperature-sensitive elements, glass or ceramic parts, or electronic components.

Welding in vacuum enhances the integrity of the components; also, the beam exerts no disturbing mechanical force during welding.

Evacuated containers can be produced without the need for post-weld evacuation.

The mass of material in instruments and other precision components is usually small and therefore of low thermal capacity. Thus, when conventional welding methods are used, such as arc or plasma, copper heat sinks or water-cooling techniques are used to limit the total temperature rise in the component, and to confine the thermal effects to a narrow zone around the weld line. The high power density of the electron beam eliminates the need for such devices; often the thermal capacity of the holding fixture is adequate. Many instruments rely on the elastic forces of springs. With EBW it is possible in certain spring materials to harden the spring and then weld it in position without affecting the temper in the critical region.

The following examples are of precision and small components being produced in batch quantities of up to one thousand parts in certain cases. They all have been successfully tested and have performed reliably in service.

Instrumentation capsules

In the design of instruments for measuring pressure or load, sensing devices are used whose movement enables a mechanical or electrical signal to be generated for the purpose of display or control. The sensing device is normally a single diaphragm or a combination of diaphragms welded together to form capsules, or one-piece hydraulically formed bellows. These sensors are either clamped or welded to retainers or other parts of the instrument.

The alloys most frequently used in the manufacture of capsules are beryllium/copper, nickel/copper, nickel/chromium/iron, nickel/iron 'constant modulus' alloy, or stainless steels of various types. The individual diaphragms are invariably thin, 0·005in. (0·125mm) being most common. In a welded capsule assembly, it can be appreciated that these flexible components may distort during welding if the heat input is high, resulting in unsatisfactory performance. Additionally, joint integrity has to be high because of the high stresses which may be encountered in operation; these are often of a cyclic nature, thus fatigue-stressing the joint. When the instrument is used for sensing pressure, leakage testing is applied to the weld, often to mass-spectrometer standards.

Argon-arc welding has been used successfully for a number of years, but it has been found that EBW not only produces joints with improved strength and leak characteristics, but also confers additional design flexibility. As an Example of this advantage two stainless steel diaphragms 0·006in. (0·15mm) thick have been welded back-to-back at a diameter of 0·5in. (12·7mm) by inserting a washer of the same material and thickness as the diaphragms between them. All three items were welded together by positioning the beam through the gap and using the washer as a filler. It is worth noticing that the diaphragms were spaced only 0·063in. (1·6mm) apart at their periphery, the diameter of which was 1·5in. (38·1mm). Such an operation is clearly impossible to perform by conventional means. The welding parameters were:

Accelerating voltage (kV)	*Beam current (mA)*	*Welding speed*
17	2	8in/min (3·4mm/s)

A more specialized aspect of instrumentation is concerned with the measurement of pressure in absolute quantities. This is often done by using an evacuated capsule or aneroid. The external pressure causes an axial displacement whose magnitude is related to the absolute pressure acting on the outer surface of the aneroid, in contrast with open capsules which indicate pressure differentials. Conventionally, the aneroid is evacuated through a small-bore tube after joining the separate components; the tube is then either cold-welded or soldered when evacuation is complete. Thus *three* types of joint are required: those making up the aneroid, joining the aneroid to the evacuating tube, and sealing the tube. Each of these is a potential source of leakage.

A residual pressure of 10^{-1} torr inside the capsule has been shown to keep the temperature coefficient of sensitivity at a tolerably low level. This residual pressure can be obtained when a single circumferential weld is produced on a pair of diaphragms using EBW at a chamber pressure of 10^{-4} torr. The two diaphragms are held together with the least loading possible during evacuation, and, as long as the diaphragms are not pierced, the required residual pressure is found to exist in the interior. Alternatively, the two diaphragms are separated during early evacuation of the chamber and then abutted by remote control at the desired intermediate pressure. Evacuation of the chamber then proceeds to the operating level of 10^{-4} torr.

Fig. 7.24 Aneroid capsule (*Courtesy Appleby and Ireland Ltd*)

The aneroid capsule shown in Fig. 7.24 is made from 0·006in. (0·15mm) beryllium/copper and edge-welded with the material in its hardened condition. Evacuation was naturally effected during welding. Experience accumulated after welding large numbers of these capsules has shown that a weld joint in such a difficult-to-weld

Fig. 7.25 Differential pressure cell (*Courtesy Appleby and Ireland Ltd*)

material can be produced with consistently higher integrity when EBW is used than with the conventional soft-soldering method. The welding parameters were:

Accelerating voltage (kV)	*Beam current (mA)*	*Welding speed*
12	3	12in/min (5mm/s)

When completed, the capsule has to be joined to instrument bodies and EBW has been successfully used for this task. Figure 7.25 illustrates a differential pressure cell revealing the capsule unit. In this example, a further operation was performed, namely welding the end-caps in position. The weld penetration was 0·3125in. (7·9mm), thus affording complete weld containment against a static pressure of 4000lbf/in^2 (27·5MN/m^2). The final welds were made with the instrument in an otherwise fully finished condition, including painting, without any damage being suffered by the delicate electrical components of the instrument. The welding parameters of the end-cap were:

Accelerating voltage (kV)	*Beam current (mA)*	*Welding speed*
27·5	96	45in/min (18·8mm/s)

The examples quoted so far illustrate the relevance of EBW to the more specialized fields of pressure-measuring instruments. However, they are typical of the trend towards the requirement for more reliability and higher temperature instrumentation. The concept of an all-welded construction affords improved positional stability of the various components of an instrument, rather than improved performance, although the latter is naturally indirectly enhanced by dimensional stability. Unfortunately, since brass cannot be welded, such an approach involves the elimination of one of the most commonly used materials in instrument construction. Other materials such as stainless steel, aluminium alloys and refractory metals have come into more general use during the last few years, a trend which will undoubtedly continue.

Proximity of weld to sensitive instrument components

It has already been mentioned that the localized, low heat input EB weld is particularly useful when the welding operation has to be carried out close to a sensitive part of the instrument. The following examples illustrate some applications.

RADIATION GAUGE. It is well known that, when polyethylene is exposed to ionizing radiation, hydrogen is evolved; the heavier the radiation dose, the greater the quantity of hydrogen generated. This principle has been applied to the design of the sensing device shown in Fig. 7.26. The assembly consists of a stainless steel tube lin. (25·4mm) dia., 0·062in. (1·6mm) wall thickness, which is first closed at one end by electron-beam welding a plug to the tube. The tube is then filled with polyethylene foil and a second plug, which carries a 12ft (3·6m), 0·125in. (3·2mm) OD copper coil, is welded in position by electron beam. After installation in the area subject to radiation, the other end of the coil is connected to a tube gauge which monitors the pressure rise as a result of evolution of hydrogen. The welding of the second plug to the tube is carried out in close proximity to the polyethylene foil and it is reported that 'electron-beam welding was adopted to reduce the incidence of leakage and, being a local

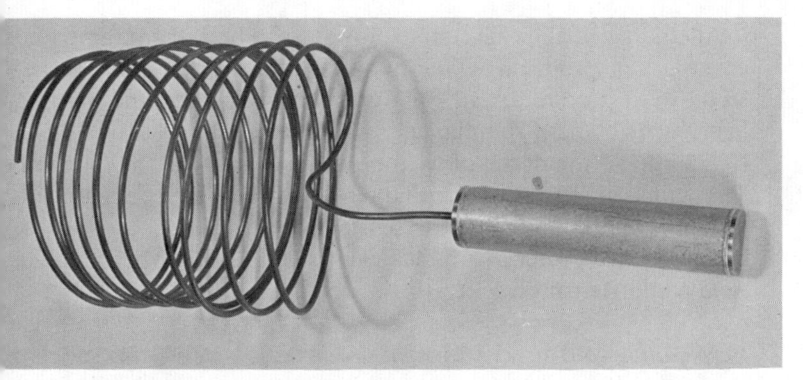

Fig. 7.26 Sensing head of radiation gauge (*Courtesy Science and Research Council, Rutherford High Energy Laboratory*)

weld process, it avoids the thermal degradation of the polyethylene which occurs with normal arc welding techniques'.[1]

ACCELEROMETER. The accelerometers shown in Fig. 7.27 were required to withstand very high overloads at some time during their service life. Stud fixing is usually used in the construction of these units, but under conditions of overload the stud stretched beyond acceptable limits. This would normally have necessitated the design and manufacture of special transducers with larger studs to accommodate the higher forces.

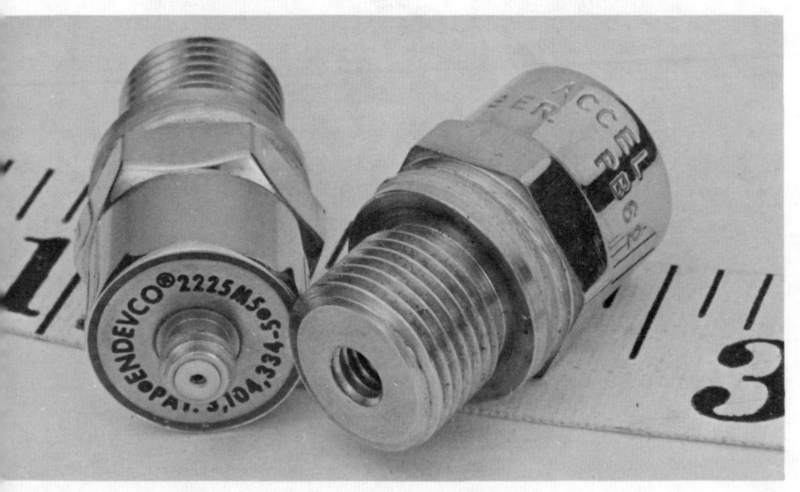

Fig. 7.27 Accelerometer (*Courtesy Endevco Ltd*)

Alternatively, and this course proved to be both more rapid and economic in producing a more suitable instrument, only larger studs were produced and electron-beam welded to the existing instrument bodies. The diameter of the weld was just over 0·5in. (12·7mm) and the depth of weld penetration was 0·125in. (3·2mm). The main point to be noted is that the weld was located only 0·125in. (3·2mm) from the piezoelectric sensing crystal.

The welded material was stainless steel (18/8/1 type) and the weld parameters used were:

Accelerating voltage (kV)	*Beam current (mA)*	*Welding speed*
19	76	75in/min (31mm/s)

PLUG EXTENSION. The plug shown in Fig. 7.28 is used in a vacuum application and the design involves the use of glass insulation in the plug. An extension to the plug was required which was achieved by electron-beam welding a stainless steel tube 0·062in. (1·6mm) thick to the main body of the plug. This was readily achieved and consistently leak-tight extension plugs have been produced. It is worth noting that the weld was carried out only 0·08in. (2mm) from the

Fig. 7.28 Plug extension (*Courtesy Cannon Electric* [*Gt Britain*] *Ltd*)

glass section of the plug without cracking the glass. Joint design is critical in such an application since weld shrinkage strains, if exerted in an undesirable direction, may cause the glass to fracture. A circumferential weld was found to be more successful than a planetary weld for this reason (see chapter 6). The weld parameters used were:

Accelerating voltage (kV)	*Beam current (mA)*	*Welding speed*
23	30	138in/min (72 mm/s)

Welding with a low power density beam

The high power density beam used in producing characteristically narrow EB welds requires closely butted faces. Sometimes joint faces cannot be produced with the required close fit and a wider, less intense beam becomes more appropriate. A similar effect can be produced by welding at lower beam power and low welding speeds. This combination generally results in a smooth upper bead which may be desirable for aesthetic appeal. Again, the wider weld produced by a low power density beam occasionally proves to be an advantage since more control can be exercised over the weld depth when welding thin sections. The following example illustrates some of the points raised above.

NIOBIUM TUBE. The requirement is to weld a cap to a 0·125in. (3·2mm) tube whose wall thickness is 0·010in. (0·25mm) (Fig. 7.29). Both parts are made of niobium which, because of its high reactivity, benefits from welding in the vacuum environment of EBW. It was an essential requirement that the weld joint should be leak-tight to mass-spectrometer standards, and also retain good ductility close to the joint.

The following low-power/low-speed settings were used to produce a wider weld, thus exercising greater control over the weld depth, where full penetration of the thin tube wall would be an embarrassment:

Accelerating voltage (kV)	*Beam current (mA)*	*Welding speed*
24	11	14in/min (5·8mm/s)

This unit is used in the manufacture of sodium high-pressure lamps.

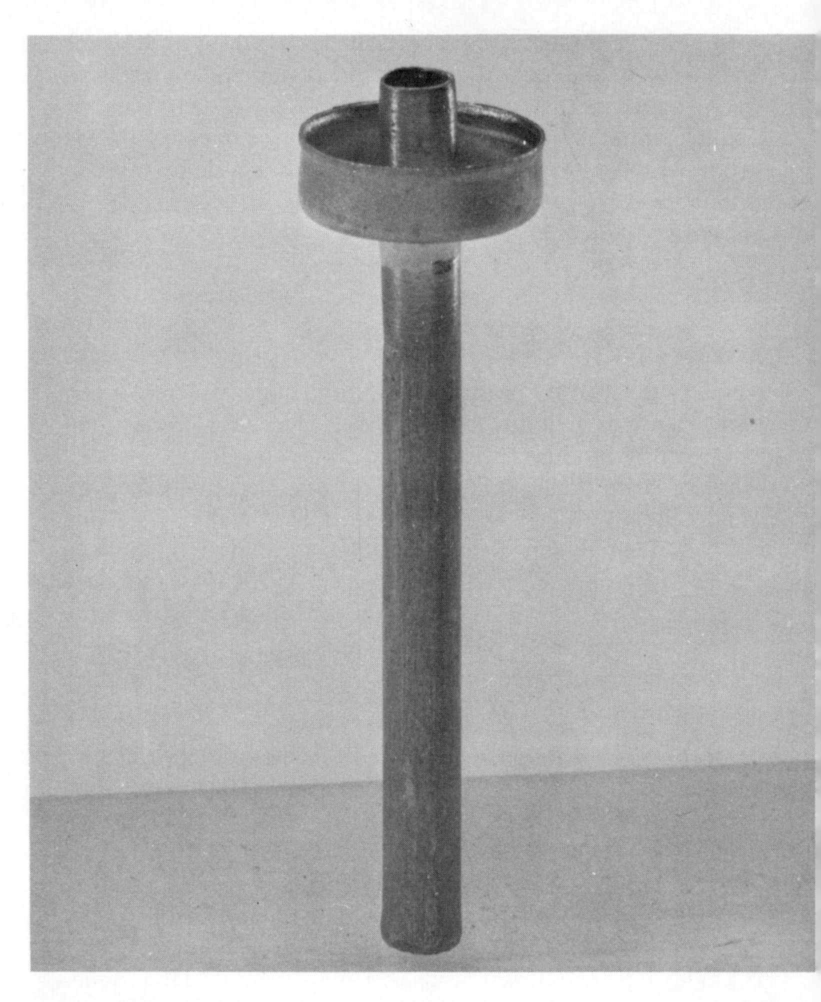

Fig. 7.29 Niobium tube used in sodium lamps (*Courtesy British Lighting Industries Ltd*)

FILTER ASSEMBLY. Porous sheet material which can be produced by sintering a number of layers of woven wire is used in the manufacture of filters. An example is shown in Fig. 7.30 where the material was sheared from strip and rolled to form a 0·625in. (15·9mm) dia. tube. This was welded along the seam and at each end to the main filter body.

Fig. 7.30 Filter assembly (*Courtesy Clifford Edwards Ltd*)

It was not possible to produce a satisfactory weld in this type of material when a sharply focused beam was used. However, by defocusing the beam by a controlled amount and operating at low total beam power and welding speed, satisfactory welds were produced. The defocusing was achieved by decreasing the current in the focusing coil by 10 per cent. Under the selected weld conditions, the beam does not fully penetrate the porous sheet; the lower layers are melted by thermal conduction and the surface tension of the molten material is adequate in preventing 'drop-through'. It is interesting to note that, although the electrons are travelling at high speed when they impinge on the molten material, the force exerted is too small to have an adverse effect on the equilibrium of the molten pool.

The collapse of the wire mesh into solid material resulting from melting produces appreciable channelling at the upper bead, but, since the joint in the filter described was not highly stressed, it was considered unnecessary to add filler material. It must be recorded that seam welds in porous sheet can be produced by argon arc; however, the advantage of using EBW lies in the cleanness of the resulting product and general reduction in scrap rate.

Precision welds

High positional accuracy is demanded in many modern instruments. This requirement can often be met by EBW and the following examples are given to illustrate this aspect of the process.

Fig. 7.31 Part of an infra-red radiometer (*Courtesy Elliott Brothers [London] Ltd and the Universities of Oxford and Reading*)

INFRA-RED RADIOMETER. The bellows shown in Fig. 7.31 are not, despite appearances, the normal type of pressure-responsive element. The convoluted tube of 1·5in. (38·1mm) dia. was manufactured by the electrodeposition of nickel to a thickness of 0·020in. (0·5mm). Solid end fittings were machined from pure nickel and welded to the electroformed tubes. The unit is used in an infra-red radiometer which measures temperature profiles in the upper atmosphere as part of the instrumentation complex of a Nimbus-D satellite.

Dimensional accuracy was important in the manufacture of this unit and the length and parallelism of two registers in each of the two end fittings had to be held within 0·002in. (0·05mm). Additionally, the joints had to be leak-tight since the units were to be filled with carbon dioxide, thus subjecting them to a continuous pressure differential of one atmosphere.

Experiments were carried out to reveal the magnitude of weld shrinkage, which was in the order of 0·002in. (0·05mm) for each joint. The bellows were therefore machined overlength to compensate for shrink displacement. Very consistent results were obtained at the following weld settings:

Accelerating voltage (kV)	*Beam current (mA)*	*Welding speed*
20	9	70in/min (26mm/s)

Fig. 7.32 High-temperature extensometer (*Courtesy British Aircraft Corporation*)

HIGH-TEMPERATURE EXTENSOMETER. An example of an all-welded instrument construction is the attachment gauge shown in Fig. 7.32. The gauge is used in strain measurement on a 1in. (25·4mm) gauge length at temperatures up to 1000°C. The extensometer operates by allowing the change of length to be measured to affect the resonant frequency of a coaxial cavity.

The 0·010in. (0·25mm) thick stainless steel bellows welded to each end of the 1·333in. (38·87mm) square blocks is used as a flexible connection piece. The small capacitor gap is fixed before the end piece is finally welded on all four sides. In this application, minimum distortion is the prime requirement.

GEAR WELDING. A small batch of twenty gears (Fig. 7.33) was required as replacements in vintage Alfa-Romeo sports cars. Although the teeth could have been cut by 'shaping', a heavy outlay on special tooling would have made the cost for such a small quantity rather excessive. Manufacturing the flange separately and electron-beam welding it into position following a relatively simple hobbing operation reduced the cost to an acceptable level without compromising

Fig. 7.33 Gear weld (*Courtesy Le Clair Precision Ltd*)

the accuracy of the unit. The gears were made of a 3 per cent Ni-Cr case-hardening steel 2in. (50·8mm) dia. with a weld penetration of 0·16in. (4mm) which was achieved at the following settings:

Accelerating voltage (kV)	*Beam current (mA)*	*Welding speed*
21·5	60	55in/min (23mm/s)

This has been only a limited selection of precision components manufactured by electron beam. Many others have been successfully produced, including ceramic/metal connectors, the fixing of small bearings into housings, thermocouples, small tube work in many forms, and the sealing of transducers in aluminium alloys. The electron-beam 'braze' is a particularly convenient method for obtaining leak-tight joints with good resistance to thermal shock. Here, the two materials to be joined may have widely different melting points or thermal conductivity; one alloy is melted and wets the surface of the second, thus producing the required joint. Suitable combinations include stainless steel to molybdenum, tantalum, and tungsten and these combinations have been used in ion engines for space propulsion. A number of transducer connections have been made between copper capillary tubes and aluminium alloy end-caps. Leak tightness to mass-spectrometer standards has been maintained even after repeated immersion in liquid nitrogen from ambient temperature. It is, however, important that the design confers adequate support to the flexing component since the joints produced may be of low mechanical strength.

Experience over a number of years in the electron-beam welding of instrument parts and other small components has shown, time and again, that the design must be worked out specifically for EBW if the full technical and cost benefits of the technique are to be achieved.

Deep-penetration welding

The mechanism of deep-penetration welding was described in chapter 3, where it was shown that this capability is primarily a result of the formation of a capillary by the intense beam. The last few years have seen the maximum power of commercially available EBW guns increase from 6kW to 50kW. With the higher power guns, weld penetrations of 3in. (76mm) in stainless steel have been achieved at sufficiently high speeds to produce narrow welds with a depth-to-width ratio of the order of 10:1. Such penetration cannot be achieved by any other heat source available at present. The ability of an electron beam to weld thick sections in a single pass without recourse to preweld preparation must be considered one of its more important assets, and one which industry at large will rapidly adopt as the potential in this important field becomes more widely known.

The main advantage of deep-penetration welding by the electron beam is the extremely low heat input. If a comparison between the various fusion-welding processes is to be made, some yardstick must be established by which the heat input per unit length can be measured for a standard penetration. It is possible to measure the width of the welds themselves, thus comparing the volume of metal melted during welding. This would give a useful indication, particularly as it is generally accepted that a narrow weld is beneficial in that the cast metal in the weld would then be subjected to a nearly uniaxial stress when the structure is loaded. However, the width of the HAZ is often as important a factor as the width of the weld, because of grain growth and metallurgical transformation. It is perhaps more appropriate to measure the heat *input* to the weld rather than the resultant effects of fusion and transformation.

A measure of the heat input per unit length of weld can be readily calculated from the main parameters of the welding operation, i.e., the power used during welding and the welding speed. Thus,

$$\frac{\text{power}}{\text{speed}} = \frac{\text{energy}}{\text{unit length}}$$

For example, a weld was performed with a beam of 30kV and 200mA at a speed of 40in/min (17mm/s). Thus,

$$\begin{aligned}\text{power of beam} &= 30000\text{V} \times 0{\cdot}200\text{A}\\ &= 6000\text{W}\\ &= 6000\text{J/s}\\ &= 6\text{kJ/s}\\ \text{welding speed} &= 40\text{in/min}\\ &= \tfrac{2}{3}\text{in/s (17mm/s)}\end{aligned}$$

$$\text{Hence, energy per unit length} = \frac{6}{\frac{2}{3}}$$
$$= 9\text{kJ/in. } (0{\cdot}35\text{kJ/mm})$$

Simply, therefore,

$$\frac{\text{energy per unit length}}{(\text{kJ/in. or kJ/mm})} = \frac{\text{power (kW)}}{\text{speed (in/s or mm/s)}}$$

The above means of comparison has been applied in two cases. In the first, EBW is compared with TIG-welding for $\frac{1}{2}$in. (12·7mm) thick stainless steel. In the second example, EBW is compared with TIG-welding of a $\frac{1}{2}$in. (12·7mm) aluminium alloy 2219. Cross-sections of all four welds are shown in Fig. 7.34, which also includes the weld parameters used. It can be seen that, for the stainless steel, the heat input for EBW is one-tenth that for TIG-welding. The ratio is even greater for the aluminium weld.

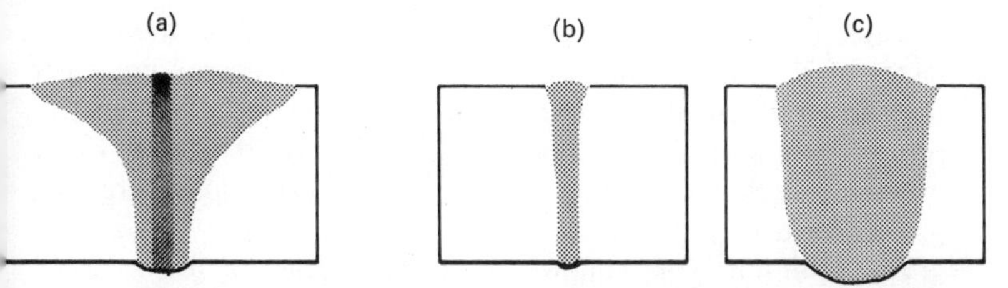

Fig. 7.34 Comparison of EBW with arc welding: (a) electron-beam weld superimposed on a TIG-weld, (b) electron-beam weld in aluminium alloy, and (c) TIG-weld in same material and gauges as (b) (*Courtesy Sciaky Electric Welding Machines Ltd*)

The energy input per unit length can vary depending on the welding parameters selected. It will be appreciated that the same penetration can be achieved by more than one combination of welding parameters, each combination yielding its own value of energy input per unit length. For example, a decrease in beam power will require a corresponding decrease in welding speed to achieve the same penetration. Again, a beam with a lower power density will require a greater total power for the same penetration at the same welding speed. In the example shown below, the same beam power and power density were used in both tests for welding 6061-T4 aluminium alloy.

	EB	TIG	EB	1st pass	2nd pass (with wire feed)	Total
Voltage	30×10^3	28·7	30×10^3	11·7	13·0	—
Current (A)	225×10^{-3}	360	200×10^{-3}	270	270	—
Welding speed in						
in/min	37	6·5	95	6·5	7·5	—
(mm/s)	(15·7)	(2·75)	(40·2)	(2·75)	(3·17)	—
Power (kW)	6·75	10·3	6	3·2	3·5	6·7
Energy in kJ/in.	9·1	95·5	3·8	29·6	28·0	57·6
(kJ/mm)	(0·36)	(3·76)	(0·15)	(1·17)	(1·10)	(2·26)

	Weld No. 1	*Weld No. 2*
Beam power (kW)	12·98	12·98
Welding speed (in/min) (mm/s)	60 (25·4)	20 (8·5)
Penetration (in.) (mm)	1·187 (30·1)	1·5 (38·1)
Energy input (kJ/in.) (kJ/mm)	13 (0·51)	39 (1·53)

Thus, it can be seen that a decrease in the welding speed by a factor of 3, which amounts to an increase in the energy input per unit length by the same factor, brings about an increase in penetration of only 20 per cent. Heat losses obviously increase at lower welding speeds, and it is important to note that such losses are even greater for deep-penetration welds.

The optimum combination of weld parameters for a deep-penetration weld with the minimum of heat input is one that combines high power density with high total beam power to produce welds at the highest possible speed. Thus, the advantages of powerful guns are being increasingly appreciated by many users, particularly for the fabrication of heavy structures.

The initial and quite considerable problem to be overcome in the application of deep-penetration welding is the development of the weld setting that can be expected to produce consistent weld geometry under production conditions, and with any number of welding operators. Such non-critical weld settings are naturally preferred because of the greater tolerances they afford. In comparatively thin sections, up to say 0·25in. (6mm) thick, good repeatability of results can be consistently maintained by the simple expediency of utilizing high beam power coupled with high welding speed. For example, when welding a section thickness of 0·25in. (6mm) in a titanium alloy, it would be advisable to use 5 or 6kW of beam power at a welding speed of 120in/min (50mm/s). By this means, not only is the weld setting non-critical, but underbead geometry can be sufficiently controlled to produce good geometry free from undercutting or other stress-raisers.

The implication not as yet fully explored is whether, as sectional thickness increases, correspondingly high welding powers and high welding speeds would continue to produce acceptable underbead geometry. Indeed, the difficulties involved in the production of a good underbead probably constitute the greatest single problem in deep-penetration welding.

A single square abutment can be electron-beam welded at slow welding speeds in certain steels and in quite heavy sections producing full penetration. However, top surface concavity is produced which, if unacceptable, has to be removed by machining. The situation at the lower surface is even more difficult since the underbead is then saw-toothed and very sharp in profile. Although this may be acceptable for many applications, the underbead can be removed as long as machining access is available. A mere witness of penetration is not recommended since the penetration will often be intermittent with some areas suffering from a lack of root fusion. If slower welding speeds have to be adopted because of lack of adequate beam power or special metallurgical requirements, resort can be made to the use of a backing support to the abutment which is penetrated during welding and subsequently removed, leaving behind it a lower bead that can be inspected and is free of the saw-toothed sharp profile. More work is required, however, to explore fully the concept of high welding powers and speeds which may provide a solution to the problem of underbead geometry in deep-penetration welds.

Recent developments in automatic wire feed may provide a more positive means of overcoming the underbead problem and the difficulties associated with producing deep welds with relatively low-power equipment. The deep joint may be made by locating on a small thickness at the root, allowing a gap between the remaining interface areas as shown in Fig. 7.35. The thickness of metal at the root could be minimal, thus allowing for close control of the underbead by adopting the high-power/high-speed concept discussed previously. After the thin location has been joined, the gap is filled by the fusion of the wire and both faces of the joint.

Two further advantages arise from the use of this filler wire technique. In the first instance, the heat input per pass will be relatively low and the weld parameters can be adjusted so that a significant proportion of the heat is used in heating up the joint walls locally. This affords a means of controlling the cooling rate of the fusion zone, and hence may provide an answer to the problem of the high hardness associated with many electron-beam welds in ferrous material which is caused by rapid cooling from the melting point. Secondly, weld metal composition can be modified by careful selection of filler wire to provide optimum physical and mechanical properties in the joint.

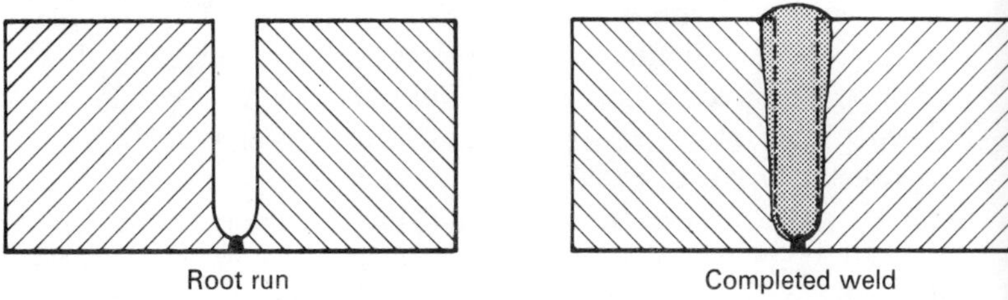

Fig. 7.35 Section through joint designed for wire-feed welding

The extension of EBW techniques to heavy sections has been applied in three major industries in the United Kingdom: aero engine, power generation, and atomic energy. Some of the more significant applications will now be examined.

The aero-engine industry

Considerable experience has been gained within the aero-engine industry since the inception of EBW and, although much of this experience is on material thicknesses below 0·25in. (6·4mm), valuable development is currently being undertaken to establish the potential for joint sections in the region of 0·5in. (13mm).

WELDS IN IMI684. The weld shown in Fig. 7.36 is fully penetrating through a 0·43in. (11mm) section of IMI684 titanium alloy. This is one of the more recent high-strength alloys intended for service at elevated temperatures, the main alloying constituents being 6 per cent Al and 5 per cent Zr. The application is a rotating part demanding a defect-free weld and a geometry which rules out any postweld machining of the underbead, thus requiring a weld geometry free from typical underbead faults. The top surface of the weld was to be machined in any case, but, as the sectional view shows, undercutting was at a minimum.

Fig. 7.36 Electron-beam weld in IMI684 titanium alloy (*Courtesy Rolls-Royce Ltd*)

Reference has been made to the high welding speed approach on thinner sections, i.e., below 0·25 in. (6·4mm), as being a successful device for controlling weld geometry. What was not known when the thicker section was initially attempted was whether it would be possible to maintain high welding speeds and reproduce an acceptable underbead geometry. The macrograph in Fig. 7.36 indicates that an acceptable weld setting can be achieved, and shows a desirable wine-glass shape of weld slug which is characteristic of titanium welds, and one which successfully avoids excessive distortion by comparison with the more familiar wedge-shaped weld. It is customary practice on titanium welds to assist the formation of an acceptable underbead by using a defocused beam. On thin sections, this is usually achieved by focusing some 2in. (50mm) below the work surface. On the section under discussion, the beam was focused 6in. (152mm) below the work surface. The focusing technique used was a simple one where a target block was placed 6in. (152mm) below the work on which the operator obtained a sharp focus. An alternative means could be found in defocusing the beam and noting the corresponding lens current. However, unless the lens-current meter is a precisely monitored instrument, the above method is preferred.

WELDS IN INCO 901. A 0·72in. (18·3mm) section of INCO 901 highlights some of the difficulties of welding a heavy section in a difficult material, probably with inadequate beam power.

Although a fully penetrating weld with an acceptable underbead would naturally be desirable, limited fusion-zone cracking cannot be avoided if a nickel-based alloy such as INCO 901 is welded at a high speed. The solution employed in this application was to include a backing support member and remove it subsequently to facilitate inspection of the weld. One of the hazards of a backed weld is clearly demonstrated by the photomacrograph (Fig. 7.37). The greater length of the weld slug is sensibly straight-sided and, at the point at which the backing member will be machined away, the weld is wide enough to ensure that there is no danger of a missed joint line. The oversized nail head, however, can induce HAZ cracking in some difficult materials. For the weld joint under discussion, a low welding speed assisted by the backing member produced an acceptable result.

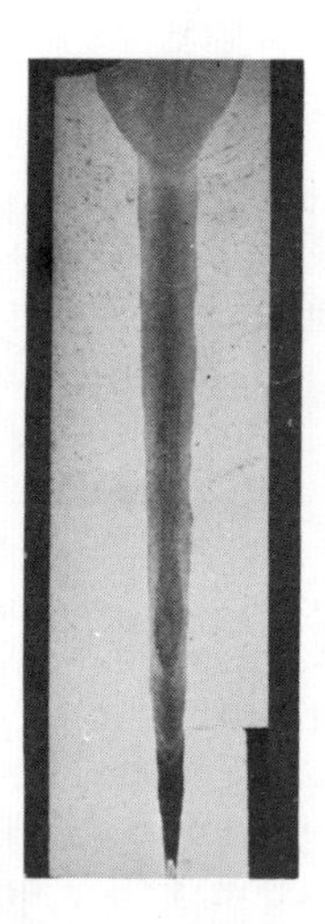

Fig. 7.37 'Backed' weld in INCO 901 (*Courtesy Rolls-Royce Ltd*)

LIGHT ALLOY WELDING. Electron-beam welding of light alloys has its own peculiar difficulties, notably the tendency to 'arc-out' in which a momentary increase of beam current is experienced owing to metal vapour travelling up the column and creating a high-voltage flash-over in the gun. The effect on the weld can be serious and has the appearance of a localized weld crater. Many precautions can be taken against this phenomenon: welding at a reasonable beam throw, i.e., 12in. (300mm) or more; employing a bent gun column; ensuring a completely clean gun before welding, and, perhaps the most positive advance in this field, differentially pumping the electron column and welding in a soft vacuum chamber environment. Welding at the higher chamber pressure of 10^{-2} torr say, is unquestionably an effective means of combating the arc-out problem. Indeed, the hard vacuum which has then to be maintained in the column by separate pumping ensures a clean gun over long operational periods.

The problem of porosity in light-alloy welding is similar to that encountered with titanium fusion-welding. In both cases, porosity can be virtually eliminated if welding at high speeds. There is ample evidence that at surface-welding speeds in excess of 100in/min (42mm/s), porosity is no longer a problem even in alloys which exhibit gross porosity when welded by conventional techniques or at low speed by EBW.

Two examples of fabrications made from light alloys will now be examined. The materials involved in the first example were a cast and a forged light alloy of the 6 per cent copper-bearing precipitation-hardening type. Both materials were welded in a fully heat-treated condition and aged only after welding. The photomicrograph in Fig. 7.38 shows a bead-on-plate run on the cast material and on which the welding parameters were determined for a depth of penetration equal to a joint thickness of 0·53in. (13·5mm), plus some penetration into a backing member, totalling 0·75in. (19mm). Although there is some porosity in the weld slug in the cast material, the as-forged stock was completely free of it, even at the low welding speed imposed by limitations on beam power. Even in the cast alloy, some acceptable reduction in the incidence of porosity was achieved by using beam-circle generation. Although it is appreciated that a more powerful gun used for welding at high speed could improve this weld, the setting has proved to be repeatable and the component was free from distortion.

The second example of deep penetration welding in light alloy materials is the fabrication of a generator casing in a magnesium alloy containing 1·5 per cent Zn, 0·7 per cent Zr. The casing is made from two concentric cylinders shrunk together to give a spiral internal cooling passage and welded at each end (Fig. 7.39). Preliminary experiments with this material indicated that, at the low speeds imposed by the limited power available from the equipment, considerable overheating took place with accompanying distortion.

Fig. 7.38 Bead-on-plate run in cast aluminium alloy (*Courtesy Rolls-Royce Ltd*)

It was also found that the penetration increased considerably as the cylinders became hot by conduction; eventually, all control over weld depth was lost. Arc-out of the gun presented additional difficulties. Previous experience had indicated the usefulness of beam pulsing in minimizing heat input to the work, and the possible use of the technique in this particular application was consequently examined. It was quickly shown that, by pulsing, not only could the deleterious thermal effects be reduced significantly but arc-out was virtually eliminated. Some porosity was unavoidable at the low welding speed, but the pores were tolerable as they were small and randomly distributed. A weld section is shown in Fig. 7.40. The only disadvantage arose from the resultant weld being only 0·02in. (0·5mm) wide which presented practical problems in the beam/joint alignment. However, with careful assembly and setting up, defects associated with lack of sidewall fusion can be avoided, and this technique has subsequently been applied to production components with considerable success.

Fig. 7.39 Part section through generator casing (*Courtesy GEC Power Engineering Ltd*)

Power-generating industry

There is little doubt that deep-penetration welding has the potential to open a new area for fabrication in which pieces at present made from expensive and heavy forgings can be manufactured by electron-beam welding smaller, more easily handled and worked sub-assemblies in an advanced stage of machining. An example of the type of forging which might be constructed in this fashion is large steam-turbine rotors. Current requirements can be met by available forging facilities, but as the rotor sizes increase, and forgings in excess of 200 tons weight are demanded, it will become desirable to use fabrication techniques. The advantages of fabrication are

Fig. 7.40 Weld section in magnesium alloy ($\times 6$) (*Courtesy GEC Power Engineering Ltd*)

numerous. The small component parts are easier to forge, to handle, and to machine. It may also be possible to weld together discs of differing chemical compositions to provide a rotor with an optimum distribution of mechanical properties.

Two deep-penetration applications have been selected to illustrate the usefulness of deep EBW in the power-generating industry. The first relates to the manufacture of erosion-resistant turbine blades. At the lower pressure stages of a conventional steam turbine, steam condensing in the form of water droplets comes into contact with the leading edges of blades at a relative speed of some 1650ft/s (500m/s) and at a temperature of 100°C. Severe erosion can take place and it is naturally desirable to protect the blade's leading edges by some means or other. One method is to attach a hard, erosion-resistant material to the blade at thc area of impact. Brazing has been used in the past, but recent development work has indicated that fusion-welding is a superior joining technique.

For economic reasons, it is preferable to attach the erosion-resistant material when the blade has almost reached the fully machined stage. Low distortion and localized heat input are two requirements that are readily met by the application of EBW.

Two erosion-resistant materials were used in the development programme. A 12 per cent Cr steel of a composition similar to that of the blading material presented no significant metallurgical problems in welding. The other, one of the Stellite range of materials, was of widely differing chemical composition from the blade, and caused some difficulties because of the difference in expansion coefficients of the two materials and the presence of porosity in the fusion zone. By rigidly clamping the assembly before welding, the effect of the difference in expansion coefficient was reduced to an acceptable level. Porosity was virtually eliminated by using a beam-spinning technique with a 2kHz frequency and 0·015in. (0·4mm) amplitude.

A blade is shown in Fig. 7.41, where it can be seen that the joint is designed to offer a linear weld seam, in some cases up to 21in.

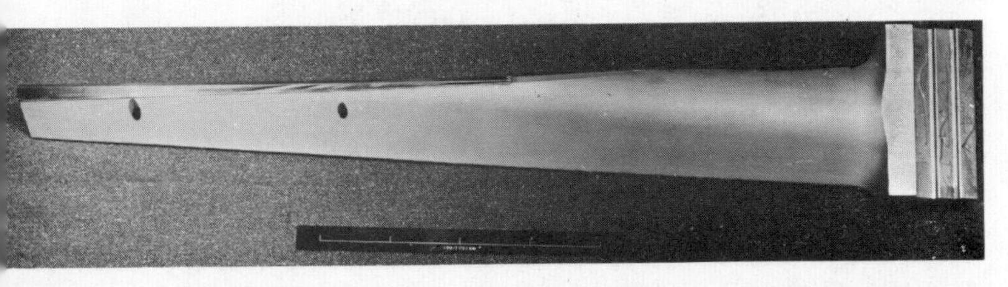

Fig. 7.41 Low-pressure steam-turbine blade (*Courtesy GEC Power Engineering Ltd*)

(535mm) long. However, because of the change in blade profile, the required depth of penetration can vary between $\frac{1}{4}$ and $\frac{5}{8}$in. (6·3–15·9mm). The power variation to accommodate the change in penetration during the welding cycle can best be produced automatically, although manual control is possible because of the relatively low welding speeds used.

The second application relates to a circulating system of a water-cooled alternator rotor, which required the manufacture of a high-pressure manifold. Attempts to fabricate this component using conventional fusion-welding techniques resulted in severe cracking of the weld metal and unacceptable distortion. Test specimens joined by EBW in the same material, a 14 per cent Cr martensitic steel, did not exhibit a tendency to cracking. Difficulty was experienced in controlling the weld profile owing to the very low welding speed of 7in/min (3mm/s). This was necessitated by the penetration requirement of 1in. (25·4mm) using equipment of maximum power of only 3kW. Further experiments showed that the use of beam pulsing produced a desirable parallel-sided weld. This introduced slight undercutting at the weld root which could not be machined out, but which was not considered detrimental under operating conditions. A comparison of the welds produced by a continuous and a pulsed electron beam is presented in Fig. 7.42. This clearly shows the poor profile and heavy weld beads resulting from the non-pulsed, slow-speed weld in this material. However, the width of the pulsed weld is only 0·03in. (0·76mm), thus demanding accurate beam/joint alignment if sidewall fusion defects are to be avoided.

The design of the manifold was consequently modified to allow for optimum welding by the electron-beam process to offer two linear butt joints about 20in. (500mm) long. These were welded successfully in a single pass. The only measurable distortion was caused by shrinkage normal to the welding direction and amounting to 0·01in. (0·25mm). The finished manifold is shown in Fig. 7.43.

Nuclear engineering

Reactor fuel-cans

The use of EBW within the United Kingdom Atomic Energy Authority has been mainly associated with fuel-element applications or with the welding of less common materials. The first EBW units were installed for evaluation purposes in 1959 and since then guns

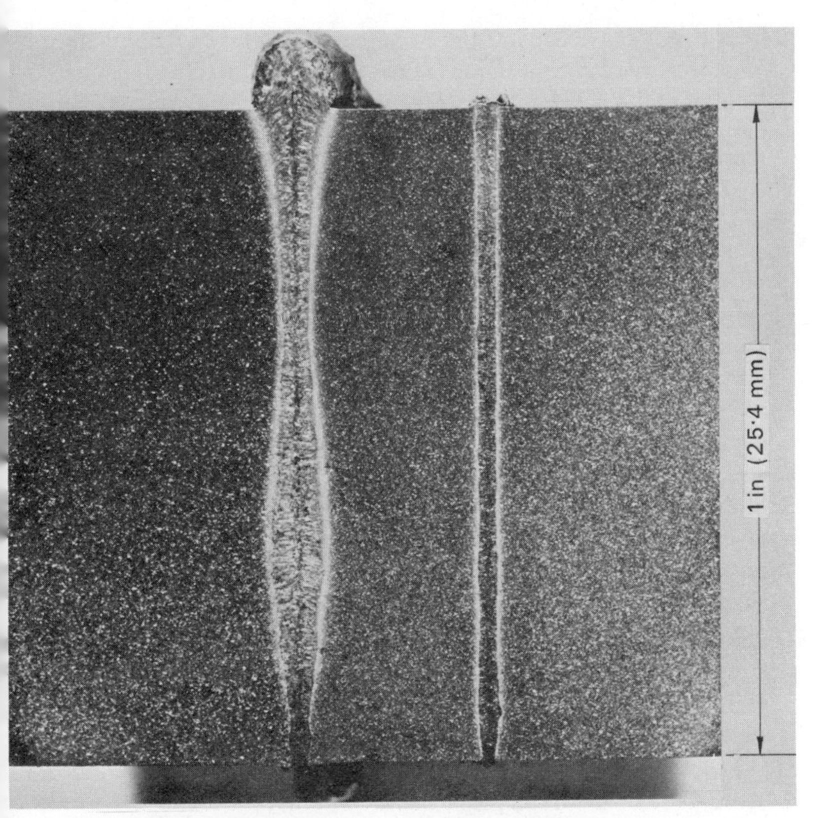

Fig. 7.42 Weld comparison between continuous and pulsed beam (*Courtesy GEC Power Engineering Ltd*)

of higher power have been installed. The machines have been used for a variety of investigations, some of which are interesting from the point of view of necessity for deep penetration.

Fig. 7.43 Water manifold (*Courtesy GEC Power Engineering Ltd*)

An alloy of particular interest in fuel-canning is Magnox AL80, a magnesium alloy containing 0·8 per cent Al. Although this material can be TIG-welded satisfactorily, the electron-beam process offered more potential advantages in providing stronger joints with greater flexibility for design. The basic design of a reactor fuel-can with an end seal is illustrated in Fig. 7.44, in which the weld is indicated in

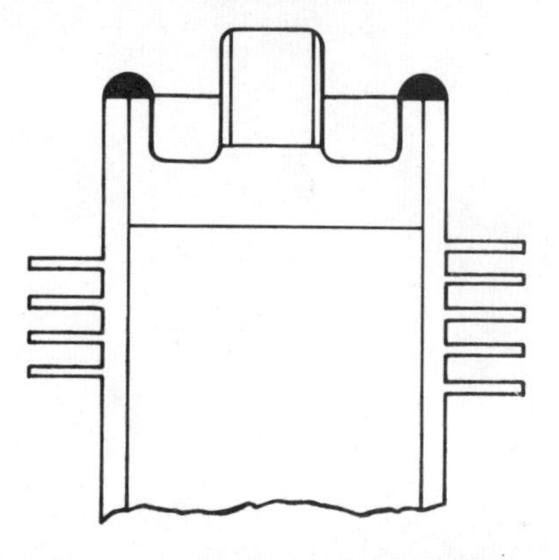

Fig. 7.44 Reactor fuel-can with end seal (*Courtesy UKAEA*)

cross-section. Experience with high-voltage equipment (150kV) shows that the rapid rate of welding obtained, combined with the small beam diameter, resulted in very little metal being evaporated during welding. Magnesium has a vapour pressure of 2·2 torr at its melting point; thus it would not appear to be particularly suited to welding in a vacuum. Trials with early low-power equipment, where welding speeds had, by necessity, to be low, showed that evaporation was indeed a serious problem which the use of beam pulsing did little to alleviate.

Continued perseverance with the process, however, resulted in the achievement of weld penetration of 0·8in. (20·3mm), with depth-to-width ratios of up to 25:1, utilizing a beam of just over 1kW. Often, the required weld was of the full-penetration type shown in Fig. 7.45, but when producing partial penetration welds, the problem of porosity was encountered. Porosity was mainly in the form of discrete vertical cylinders approximating to the weld pool diameter and occurring at almost regular intervals. If the beam was suddenly switched off on completion of a weld, a hole was left which usually constituted a leak path. Alternatively, a gradual decay of beam power, or increase in welding speed, gave a line of discrete porosity extending to the surface. In this case, a leak-tight weld would be produced but the effective leak barrier left much to be desired.

Fig. 7.45 Full penetration weld (× 10) (*Courtesy UKAEA*)

Further investigations showed that improvements could be obtained by decreasing the depth-to-width ratio; the use of low-voltage (30kV) equipment was therefore investigated. However, it was found that the reduction of the depth-to-width ratio resulted in excessive undercutting, increased spatter, and an unacceptably high rate of metal evaporation. However, the low-voltage welds exhibited less root porosity and the wider weld bead produced meant that alignment of the beam with the weld joint was less critical.

Comparison of high-voltage welds made with beam oscillation (to reduce the accuracy of alignment) with the low-voltage welds showed that, for an equivalent depth of penetration of approximately 0·4in. (10·2mm), the low-voltage weld was made at 90 per cent of the power and at 50 per cent of the speed of the high-voltage weld. Depth-to-width ratios were approximately 10:1 for low-voltage and 14:1 for high voltage. On balance, the high-voltage welds were preferred for this application since the welding process was easier to control and, a major deciding factor, far less metal evaporation was involved.

A conclusion rapidly reached during the development stages of welding Magnox was that the design of the vacuum chamber, in the region of the weld joint, had to be such that there was no restriction which would adversely affect pumping efficiency. Restrictions which might not cause any problem with materials of low vapour pressure resulted in a local concentration of magnesium vapour which effectively defocused the impinging beam and consequently affected the weld produced. An allied problem is the penetration of magnesium vapour to the upper reaches of the electron gun which can cause flash-over between the anode and bias electrode, particularly at voltages in excess of 120kV, resulting in excessive beam currents and damage to the weld.

Although root porosity could not be entirely eliminated, it could be reduced to acceptable proportions providing a product far superior to that produced by TIG-welding. As can be seen from Fig. 7.45, the weld bead grain size is extremely small, unlike that of a TIG-weld, and no HAZ can be detected. These features, together with a larger bonded area, were reflected in improved mechanical properties; the failure load of an EB welded end seal was twice that of a TIG-weld.

Electron-beam welding has been evaluated by the UKAEA for the production of reactor fuel-cans, employing unskilled labour. The machine used had fully automated welding sequences, the operator being responsible only for loading the equipment and initial alignment at the start of the weld. In this application, a straightforward replacement of TIG-welding by EBW can show a marginal economic saving in addition to a much improved product. Furthermore, by taking full advantage of the design flexibility offered by the EB process, cost savings can be made in other directions. Thus, the electron-beam welded component is potentially more competitive, on a purely economic basis, than that produced by TIG-welding.

Salvage

As engineering products continue to develop in complexity and sophistication, so does the cost of scrap. In mass-production industries, an error in manufacture can extend to a large number of parts before it is detected, with a corresponding increase in the cost of scrap. New salvage techniques are always being examined, and EBW is establishing itself as a particularly useful salvage method since it possesses certain unique features, such as:

precise control of position of beam so that it is accurately aimed at the area to be repaired;
ability to find its way through narrow gaps, reaching areas that are normally inaccessible to other heating sources;
heat input can be so small that it causes the minimum disturbance to the properties of the material and the least distortion to the part.

There are, however, one or two aspects of salvage by EBW that may restrict its application. Because of the need for welding in a vacuum chamber and the high welding speeds of EBW, the salvage operation has to be mechanized. If the weld path in a salvage operation is irregular, this may present certain difficulties, although use can be made of line-following techniques. Again, if the salvage operation requires additional material such as the filling of a crack, wire feed methods will have to be employed, again in a vacuum chamber. It is often convenient to cut out the section containing the error, welding a new part in position. The weld line can be arranged to be simple in geometry, such as a circle; also the abutment can then be made free of gaps.

Salvage techniques are also used during the period of development of a new product. Design changes are introduced as development experience accumulates. It is often more convenient to alter an existing part than to manufacture new ones. Again, damage may be sustained during service which would be more economically dealt with by repair operation. Besides cost advantages, there are also hidden benefits derived from the speed of repair as compared with the normally long cycle of manufacture of a new part.

The following are examples of salvage operations successfully carried out by EBW.

Pressure transducer

The body of the transducer, which is part of a servocontrol system, is made from an austenitic stainless steel. The transducer tube (Fig. 7.46a) contains a mass of intricate electrical circuitry which is encapsulated in resin. The final manufacturing operation involves the sealing of the end of the tube. Initially, this was carried out by using an epoxy adhesive but service experience has shown that the bond obtained was unsatisfactory. Fusion-welding was then considered and EBW was selected as it was likely to cause the minimum disturbance to the delicate electrical parts.

Figure 7.46b shows a sectional view of the part of the assembly involved in this operation. The 0·003in. (0·08mm) thick disc was secured to the transducer tube by means of light weld tacks, produced with a miniature resistance welder. The EB weld was carried out by rotating the transducer below the beam. An enlarged view of the weld produced is shown in Fig. 7.46c. The machine settings used were: 55kV, 1·5mA, 4s weld time. The welding conditions were found to be critical: a slight excess of heat produced outgassing of the potting medium. The settings were produced using a simulation of the component, the weld penetration was limited to 0·006in. (0·15mm) and the weld width to 0·022in. (0·55mm).

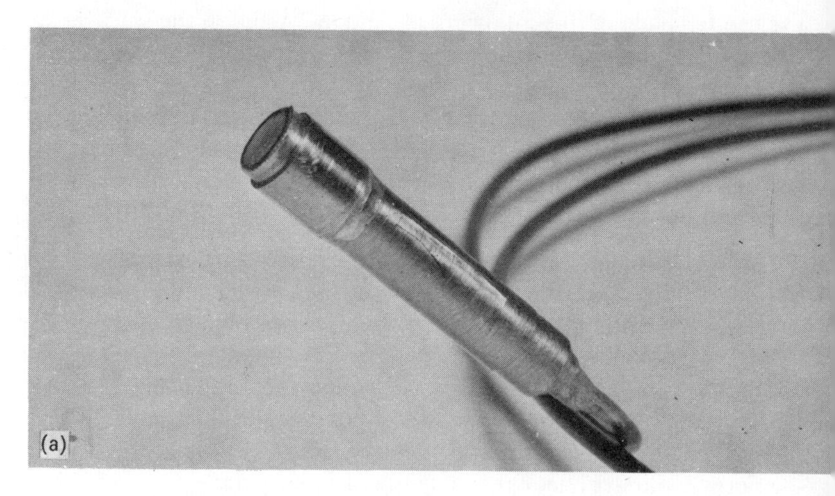

Fig. 7.46 Pressure transducer: (a) transducer tube, (b) sectional view of joint, and (c) enlarged view of weld (*Courtesy Rolls-Royce Ltd*)

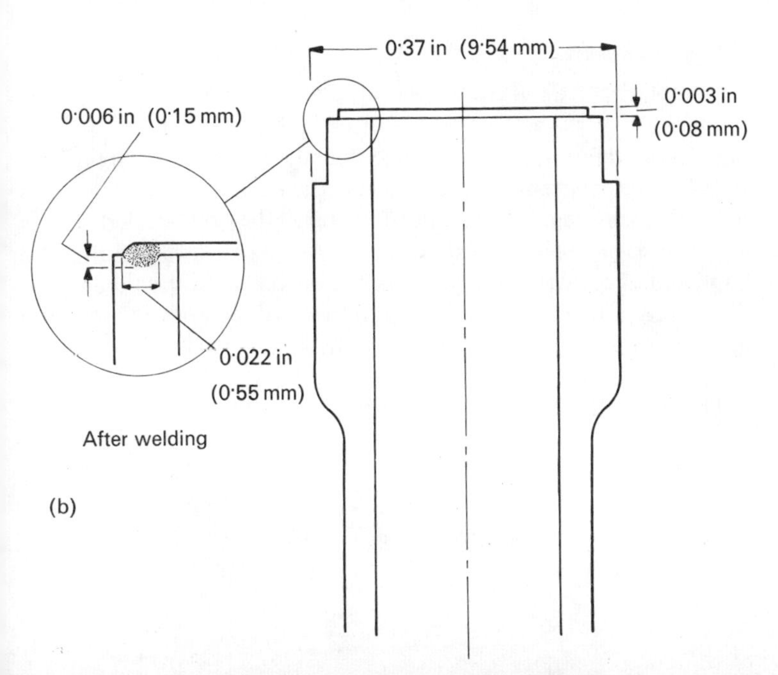

The postweld examination included radiography, leak testing, followed by functional tests, all yielding satisfactory results. It is difficult to conceive how else this joint could have been produced. Apart from the use of adhesives, which had been proved to be unsatisfactory, any other welding or soldering technique would, by the amount of heat generated, destroy the intricate components of the transducer. Electron-beam welding was found to be sufficiently precise and controllable for this exacting application.

Compressor shaft

A recent trend in axial compressor design is to produce one or more compressor discs integral with the shaft. An example is shown in Fig. 7.47a where a design change demanded a longer shaft. The forging, in a martensitic stainless steel, was then in an advanced stage of machining. It was decided to cut off the original shaft, welding a longer one in its position. Figure 7.47b shows the joint preparation used. This was locally thickened in the weld area as it was decided to machine away the weld upper and lower beads to avoid stress concentration effects, thus producing a more fatigue-resisting joint.

The first EBW operation consisted of tacking the joint at eight

Fig. 7.47 Compressor shaft: (a) disc integral with shaft, (b) joint preparation, and (c) weld overlap (*Courtesy Rolls-Royce Ltd*)

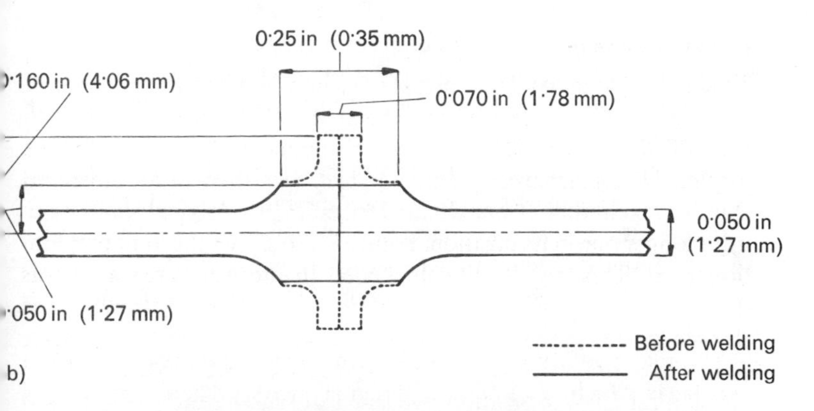

equispaced stations using the machine settings: 145kV, 14mA, 0·040in. (1mm) dia. effective beam circling, and weld duration 1s. This was followed by the full weld pass at 145kV, 14mA, and a welding speed of 24in/min (10mm/s).

The beam decay at the overlap point was produced in 10s at a constant accelerating voltage. The weld overlap is shown in Fig. 7.47c.

The weld was examined by radiography, and by magnetic and ultrasonic tests. Similar tests were carried out after machining the weld and surrounding area to final size. Dimensional changes resulting from both the welding operation and inaccuracies due to fixturing were found to be very small. Any distortion inside the bore of the shaft is particularly critical, as certain areas are fully machined before welding. Further, as the shaft rotates at high speed in service, balancing problems would be encountered in cases of distortion.

One of these shafts was electron-beam welded using the above sequence and subjected to a rotating fatigue test, which extended well beyond its expected life in service. The weld was found to be completely satisfactory.

This major rotating component is among the first to be made from two parts which are joined together by EBW. The component is highly stressed, being subjected to both static and cyclic loading. The exacting tests have proved that this method of manufacture is feasible and produces a shaft of the high integrity required for this application.

Constant-speed drive unit

A component part of a constant-speed drive unit illustrated in Fig. 7.48 is the 'wobbler' housing. The housing is machined from a solid forging in 17 per cent Cr–3 per cent Ni precipitation-hardening steel, and measures 4in. (102mm) dia. by 6in. (152mm) high. The component involves a considerable amount of machining which represents the greater proportion of the total cost.

Inspection of a batch of twenty-seven housings revealed that the holes drilled in the upper flange were wrongly positioned. Since the housings had reached an advanced stage of manufacture, consideration was given to the possibility of salvage. As plugging of the machined holes was unacceptable, it was decided to attempt to replace the complete flange. Mechanical fixing would not have provided sufficient strength and fusion-welding was then considered. Arc welding was rejected as it produced an unacceptable degree of distortion in the near-finished component. Electron-beam welding offered the attraction of high strength coupled with minimum distortion.

The drilled flange was machined off and a substitute blank ring was manufactured. A simple location recess was machined in the body of the housing on the outside diameter (Fig. 7.48a). The replacement ring was turned to give a radial fit-up with the recess with a tolerance of $-0{\cdot}000 + 0{\cdot}001$in. ($-0{\cdot}000 + 0{\cdot}025$mm). The ring

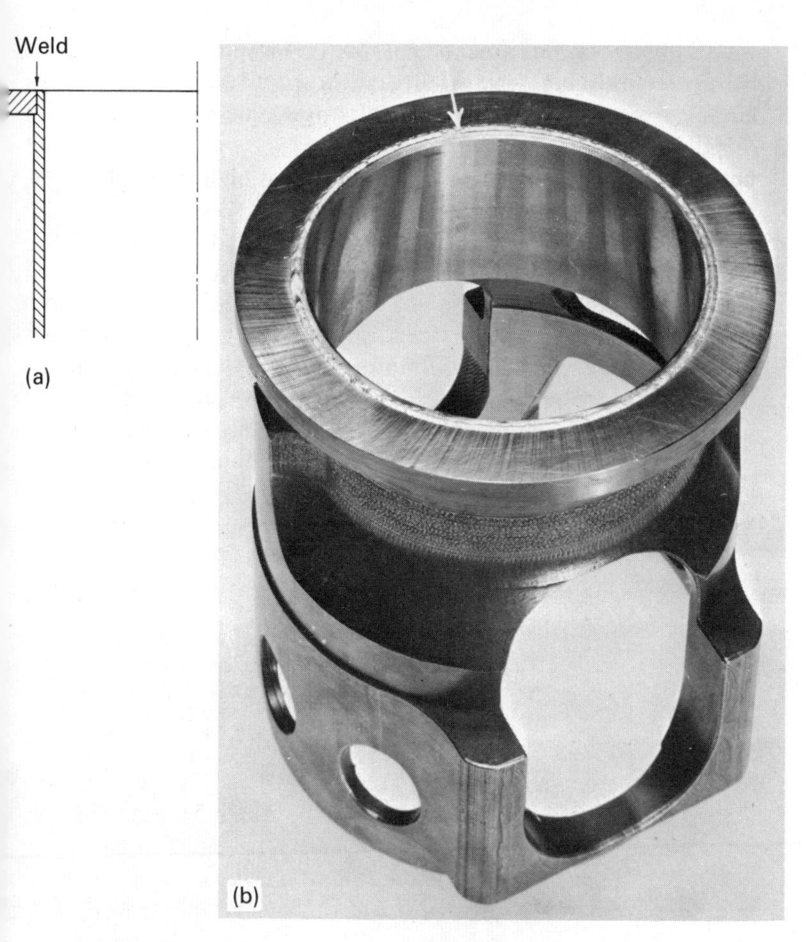

Fig. 7.48 'Wobbler' housing: (a) location recess, and (b) welded flange (*Courtesy GEC Power Engineering Ltd*)

was then electron-beam welded by means of a single pass at a speed of 45in/min (19mm/s). The beam setting of 140kV and 12mA produced full penetration of 0·25in. (6·3mm).

Non-destructive examination by ultrasonic methods revealed no defects in all but two of the twenty-seven housings. The potential defect was some lack of sidewall fusion which was remedied by applying a second weld. The distortion was well within the allowable tolerances.

Although material cost savings secured by the success of this salvage were not high, significant savings in both time and machining costs were attained. It is worth noting that the self-locating joint design made it possible to perform the welding operation without any jigging, thus reducing both the cost of salvage and the time cycle involved.

High-speed gear

This component, a 6in. (152mm) dia. gear unit in EN39, had been tested and, in the light of service performance, a design modification was introduced which called for an additional $\frac{1}{2}$in. (12·7mm) on the total length. Arduous service conditions precluded the attachment of an extension ring by mechanical methods, but a sound fusion-weld was acceptable. The gear teeth were finish-machined and distortion

Fig. 7.49 High speed gear: (a) location spigot, and (b) welded gear (*Courtesy GEC Power Engineering Ltd*)

in excess of 0·0005in. (0·0013mm) on the teeth was unacceptable. Electron-beam welding was selected as the process most likely to produce a satisfactory joint while maintaining the machined tolerances.

Examination of a test weld in EN39 showed that the weld and HAZ hardness were acceptable. No weld defects were found.

A 0·01 × 0·01in. (0·25 × 0·25mm) locating spigot was machined on the gear body and a matching location machined on a $\frac{1}{2}$in. (12·7mm) thick annulus (Fig. 7.49a). The parts were located and held in contact by means of head-and-tail stock unit. A single-pass weld which fully penetrated the annulus thickness and fused the entire spigot location into the weld was performed (Fig. 7.49b). The weld parameters used were 150kV and 28mA at a welding speed of 60in/min (25mm/s).

Non-destructive examination after welding revealed that the weld was completely sound. There was also no measurable distortion of the gear teeth.

Filter unit

The unit shown in Fig. 7.50a which is made of 18/8 stainless steel acts as a slurry filter in a turbodrill. Damage often encountered is the scoring of the ground surface owing to the presence of grit. This naturally destroys the sealing capability of the unit and, since the part was otherwise sound, a salvage procedure was called for. This would replace the metal removed by the grit, to be followed by a machining operation restoring the sealing surface to its original shape. It was particularly important for no distortion to take place during the salvage operation.

The damaged area was chemically cleaned and two half-rings were located in the 'O' ring groove close to the sealing surface to avoid damage to the groove during welding. Automatic wire feed was employed to deposit material into the damaged area. Figure 7.50b shows the unit after welding and a macrosection of the welded area is shown in Fig. 7.50c. The extra material deposited was some 0·064in. (1·6mm) which was fused into the parent material to a controlled maximum depth of 0·1in. (2·5mm). The total energy used for this operation was 0·011kWh.

Destructive tests on simulated parts were undertaken which revealed sound fusion and satisfactory metallurgical structure. Dimensional checks which were carried out showed no measurable

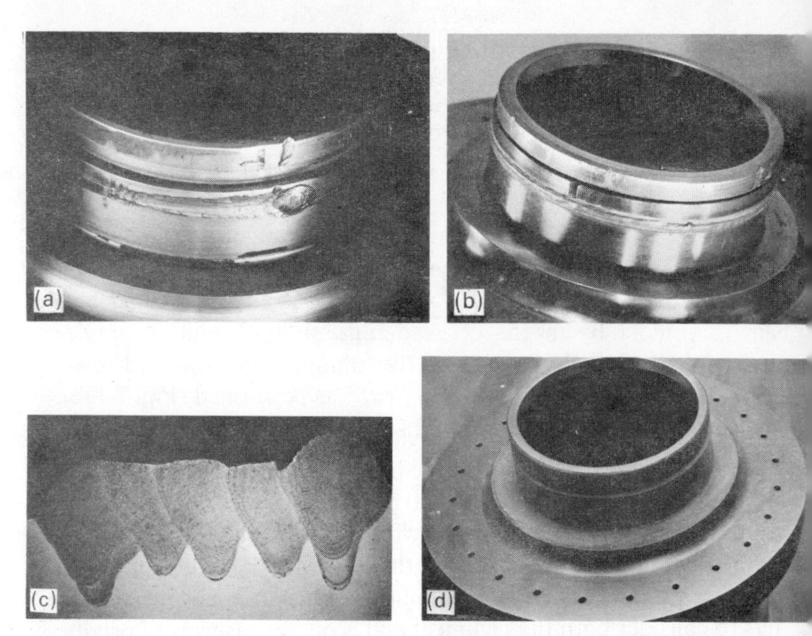

Fig. 7.50 Filter unit: (a) damaged sealing face, (b) welded area showing protective ring, (c) sectional view of deposited metal (× 5), and (d) salvaged unit (*Courtesy Rolls-Royce Ltd*)

change; this was because of the rigidity of the component itself and the low energy used in welding. The component was machined to its original dimensions (Fig. 7.50d), followed by rig and field tests. The component's performance was no different from that of a new one.

The success of this salvage was due to a combination of favourable factors: the rigidity of the part, the precision of the EBW operation, the successful development of an automatic wire feed method, and the minimal amount of energy used in performing the fusion operation.

REFERENCE

1. Morris, A., Sheldon, R., and Stapleton, G. B. Rutherford Laboratory Report No. RHEL/R 132, October 1966.

8. Production engineering

In earlier chapters, the metallurgical and design aspects of EBW have been considered. Once these have been satisfied, it is then the duty of the production engineer to examine the practicability of the application, taking into account whether the part can be produced at the right price, in the right quantity, and with the right equipment, or more probably with the equipment he has at hand. Broadly, he has to consider both the technical and economic aspects of production, optimizing them to meet specific requirements.

The production engineer has to consider the relative merits of competing processes, though sometimes there is no alternative. His task is then easier in one way, but he has to examine, probably in collaboration with a marketing colleague, whether the cost of his EB weld can be justified in relation to its value to the product. By contrast, there are applications where EBW is in direct competition with other joining methods and offers no particular technical advantage. The choice is then based simply on economic grounds, provided the required rate of production can be met by both processes. Recent trends reveal that this class of application will grow in volume, as the potentiality for EBW to confer a lower overall cost is appreciated.

The bulk of EBW application, however, falls in a class where the technique offers an improved product, although other methods may still be feasible. The improvement may be offset by an increased cost of production and then the two factors will have to be balanced one against the other. However, it is important always to compare the *total* cost of the product and not simply the cost of the welding process alone. The total cost will include such factors as the cost of the material and any variations in its specification, the preparation before welding, the heat treatment, the welding operation itself, the inspection, and, by no means least, the cost of scrap or salvage.

It is more convenient to consider the two aspects of production, technical and economic, separately. It is clear from what has been said earlier, however, that the two interact extensively and must both be taken into account before a sound basis for optimum production can be reached.

Technical considerations

Electron-beam welding has been shown to have great potential as a joining technique, the more so when full use is made of the high welding speeds available. The correct design of the parts, the proper planning of work to ensure a smooth flow into and out of the machine, the intelligent use of suitable jigs and fixtures, and regular maintenance to reduce down-time to a minimum can do much to ensure economic running of the plant.

Work handling and ancillary equipment

Work handling is just as important to the success of the application of EBW as any of the basic units of the plant. Since all manipulation of the workpiece within the vacuum chamber has to be carried out by remote control and since, in general, high welding speed will preclude manual adjustment of parameters during the welding sequence, the manipulating device once set must be capable of carrying out the complete programme. It is beyond the scope of this book to describe in detail all types of fixtures, manipulators, and other equipment connected with work handling, but the underlying features and basic principles of general tooling for EBW will be introduced.

Jigs and fixtures

Previous chapters have emphasized the necessity to maintain component parts in intimate contact so as to avoid defects through lack of fusion. Often, this problem can be reduced by simply increasing the weld width, but sometimes this is undesirable because of the introduction of increased distortion or because of the effect of the greater heat input on joint properties.

The jig in EBW therefore assumes the important role of locating the component parts, maintaining an intimate contact at the interface. It also acts as a restrainer and holder. Although distortion is low, even a very small movement of the joint interface with respect to the electron beam can result in the weld line being missed and hence lack of fusion. The most critical function of the jig is therefore to restrain the component during welding. Since, in most cases, the job and not the welding head moves during the welding cycle, the jig often acts as a holding device through which the relevant actions are transmitted. Sometimes, component parts can be made to be self-locating, as in the case of welding a disc to a tube, but, more frequently, separate locating devices have to be employed.

As the number of parts to be welded increases, the production engineer has to decide whether to adopt multiple jigging in one form or another or to use one-off jigging. This is essentially an economic choice and the economic aspects are dealt with in another part of the chapter. When using one-off fixturing, however, the evacuation time of the chamber may well represent a significant proportion of the total time cycle, especially if the chamber is large. The use of multiple fixturing, although expensive, can reduce the time spent in pumping and hence increase the overall utilization of the plant.

As an alternative to multiple fixtures, which in any case may not be a practical proposition for large parts, the concept of vacuum locks can be considered, particularly when both the evacuation and loading assembly times are long. One suggested idea is a system which comprises three chambers connected in series : one for loading, the middle one for welding, and the third for unloading. Work is loaded into the first chamber and transferred to the second, which is then isolated from the first and evacuated. The first chamber then receives the second part and can be evacuated while the first part is being aligned and welded. The third chamber is kept evacuated to receive the part after welding. Parts thus pass through the system, the centre chamber always being under vacuum so that welding can take place immediately the part is transferred to it. In principle, this system has attractions, but problems exist in work handling, in the transfer gear, and in the vacuum locks. The pump-down time of the loading and the unloading chambers must not exceed the welding time as otherwise they will hold up the flow of work. Evidently, for such a unit to function satisfactorily, some degree of automation will have to be introduced so that parts can be manipulated between chambers without the risk of accidental evacuation or air admission to the wrong compartment. This soon leads to an expensive and complex piece of plant, the design and construction of which are not to be undertaken lightly.

It may be that a considerable proportion of the total time is consumed by the assembly of the parts in the fixture. In these circumstances, the evacuation time can become insignificant and it is then clear that duplication of the fixture, single or multiple, is desirable.

Fig. 8.1 Multiple-fixture for circular welds (*Courtesy Hawker Siddeley Dynamics, Electron Beam Division*)

One fixture assembly can then be set up outside the welding chamber while another is placed inside the chamber ready for welding.

In the interests of economy, jigs and fixtures should always be as versatile as the design will allow, so that one fixture can be used for several applications of the same general nature. Circular welds can be handled by the jig shown in Fig. 8.1 which is capable of rotating fifty-six parts simultaneously. The fixture is designed so that bushes can be inserted into the holders to accommodate a range of diameters up to a maximum of 1·5in. (37·5mm) with every bush in use. Larger bushes can be used, although the number will have to be reduced. Another example in the versatility of the jig design is to use an eccentric table in which a pair of adjustable cranks is used to drive a flat table so that all points on the table describe equal circles. Such a unit, shown in Fig. 8.2, has proved to be quite useful by many users. The cranks are adjustable to produce circles

Fig. 8.2 Eccentric-drive table (*Courtesy Hawker Siddeley Dynamics, Electron Beam Division*)

in the radius range from 0 to 1in. (0–25mm), although this can be increased if necessary. Obviously, this jig eliminates the need for setting components accurately under the beam since, irrespective of the position of the workpiece on the table, a circle of the preset diameter will be described by the crank. The technique is particularly useful for welding complex parts which may otherwise be difficult to jig.

Jigs are often used to maintain the interface of the part being welded so that it coincides with the position of the focused electron beam. It is essential, therefore, that jigs should be designed and manufactured to an accuracy in keeping with that of the existing manipulating devices. If the jig is used repeatedly, it is also important that the operation should be reproducible. The absolute accuracy depends upon a large number of factors, the majority of which vary from job to job, but the tolerances can be relaxed when a wider weld is permissible.

The jig and the part having been set in the chamber, it is almost invariably necessary to adjust the manipulator to align the joint line with the electron beam. This can be tedious and time-consuming, particularly if a multiple jig is employed. To eliminate alignment procedure, it is often possible to produce jigs which index positively and locate in a fixed position in the welding chamber with the joint seam directly below the electron beam.

A possible variation of the jig principle is to regard the vacuum tank as the part which holds the work accurately for manipulation under the electron beam. Here we are obviously stretching the definition of the word jig. The implication is that the fixture is used to hold the gun in the correct relationship to the manipulator. The workpiece has still to be attached to the manipulator.

The significance of this concept is that, for an awkward or large workpiece, it may be economically sound to build a special vacuum tank which is tailored to fit the workpiece while allowing it to be manipulated as required. The combined tank-manipulator-component assembly is then brought to the gun. The possible technical advantages are, firstly, the work can be located more rigidly and more accurately under the gun. The initial cost of the vacuum enclosure can be amortized in the same way as the holding jig, being part of the jigging assembly. Secondly, the gun may be mounted in the optimum position to carry out the required weld. The possible cost advantage is that, by reducing the chamber volume to a minimum, the pump-down time is also minimized. However, this approach is feasible only when the extra cost of jigging can be amortized over a large enough number of components, and when it is a simple matter to move and to remount the electron gun and the vacuum pumping system.

Manipulators

The relative movement between the worktable, to which the part and the jig are attached, and the electron beam can be produced either by moving the electron gun or by moving the table. Many different types of equipment are now available with a wide variety of movements, but most machines are supplied with two horizontal linear motions, one along the chamber length and the other perpendicular to it, the *X* and *Y* axes respectively. There are also facilities for rotating parts about the horizontal and vertical axes. Essential features of the moving table are accuracy, close control of velocity, and reproducibility; the need for adequate load-carrying capacity is also becoming significant as more heavy parts are now being electron-beam welded.

Components calling for non-linear and complex weld patterns require special handling methods. Various techniques have been adopted for weld-programming where the operating system is either mechanical or electronic. In principle, complex horizontal joint lines can be followed by controlling the *X* and *Y* motions simultaneously; this can be achieved by setting up, outside the vacuum chamber, a pattern or template of the weld line and following it with conventional mechanical tracking equipment, relaying the resultant signals to the *X* and *Y* drives. This is not a closed-loop system, so there is no correction for errors such as those experienced, say, from the backlash inherent in mechanical linkages. Such a mechanical method is therefore not entirely suitable for use in conjunction with EBW, which requires greater accuracy in profiling. Alternatively, the pattern can be followed by using a photoelectric line-sensing device, but the limitation here is associated with the sensitivity of the photocell to detect changes from black to white. Generally, an accuracy of 0·004in. (0·1mm) can be obtained, which is adequate for most applications. Where greater accuracy and certainty are required, it is better to use a system which follows the joint interface itself, producing a signal to adjust the position of the workpiece at the point of impingment of the electron beam. Such seam-tracking systems are described in chapter 4.

As an alternative to mechanical methods, the relative movement may be accomplished by deflecting the beam magnetically. Many EBW equipments contain a facility for deflecting the beam by means of electromagnetic coils located at the base of the gun. The beam can be deflected through an angle to the axis of the gun by energizing the coils in the relevant direction. This facility can be used to deflect the beam to follow regular patterns quite easily, e.g., circles, squares, ellipses, or a complex path dictated by pattern-tracing equipment or a programme.

A complexity with magnetic deflection is that, not only must the electron beam follow the correct path, but it must also do so at the correct speed, generally a constant speed. Where the path described is a regular shape, the requirement is relatively easy to satisfy, but where it is complex, the problem can become really difficult. Caution should, however, be exercised since reproducibility of positioning may be impaired by deflection-coil hysteresis. Beam focus also deteriorates with increasing deflection angle. Also, as the deflection angle increases, the beam no longer impinges normal to the component surface and is not able to follow the weld seam in depth. This consideration is of particular importance where deep penetration is sought. In theory, this error can be corrected by using two deflection-coil systems, with the second bringing back the beam to its vertical position. The second deflection coil has to be of wide aperture and the method is applicable only when the pattern is limited to a small area.

It is perhaps appropriate to mention here that the facility to deflect the beam to follow a regular pattern can be adopted to increase the effective area of the beam by oscillating it through a small amplitude in a direction normal to the weld seam. This is not the same as using a larger diameter beam of lower intensity. An oscillated, well-focused beam has deeper penetration than one of low intensity. Also, there is a good deal of evidence that the oscillated beam can be used to minimize the porosity evident in some materials welded with a very narrow static beam. This matter is discussed in greater detail in chapter 5.

Techniques of alignment of the electron beam with the joint interface before welding have been described in an earlier chapter; the significance of maintaining this alignment during the welding sequence has also been emphasized. The high welding speeds used in EBW, seldom less than 20in/min (8mm/s), preclude manual adjustment to offset deviations from correct alignment. Automatic correction can be made by using equipment to track the seam. The weld seam, or a line parallel to it, can be followed either by electrical methods, by a scanning electron beam, or by the use of a mechanical stylus. Deviations from the seam transmit a signal back to the deflector coils or the appropriate drive where a correction is made. Standard equipment for this purpose based on the mechanical stylus principle is now commerically available (see chapter 4). Error-correction techniques can also be extended to cover lack of accuracy in original alignment of the workpiece.

Accessory drives and controls

Manipulating devices designed and built to work in a vacuum environment for EBW need to meet special requirements which include:

mechanical accuracy, rigidity, reproducibility, and freedom from backlash;
absence of components liable to produce static or dynamic magnetic fields;
absence of materials likely to produce contaminating vapours;
mechanisms which will not be clogged or jammed by weld spatter or deposited metal coatings;
use of bearings and running surfaces which do not require conventional lubricants or which can run dry;
mechanisms which work with high power efficiency, or have means to carry away the waste heat.

In the majority of commercially available equipment, drive motors and gear boxes are located outside the vacuum chamber so that the problems associated with running electric motors and lubricated parts at low pressures are overcome. Drives enter the work chamber through rotary vacuum seals. It is preferable if the motors, whether hydraulic or electric, are of a variable speed, reversible type with high acceleration to the preselected speed; this should be maintained constant throughout the welding sequence to prevent variations in weld characteristics from being introduced. It is possible to operate hydraulic motors, rams, or low-voltage d.c. motors at the pressures prevalent in EBW, but mains voltage motors must be located outside the chamber or sealed in pressure-tight containers.

To ensure adequate weld penetration, slight excess energy must be supplied to the joint and the surplus beam current may cause damage to surfaces below the weld seam. To absorb this energy, it is general practice to insert, close to the underside of the joint interface, scrap material which can be removed after welding. It is advisable to use the same material as the component, to avoid the likelihood of metallurgical contamination. The protective material can also act as an indication that full penetration has actually occurred. In certain instances, for example in welding sealed components, this protection will not be possible, but if the excess power is kept at a minimum, the consequent damage on the inside of the part may then be tolerable.

The mechanism of EBW involves the rapid flow of vapour and liquid up and down the weld cavity, which results in some material being expelled with considerable velocity from the joint line during welding. This takes the form of metal vapour, droplets, and spatter. The vapour accumulates as a deposit on surfaces unobstructed from the joint line, while the droplets and spatter ricochet inside the vacuum system, landing on most surfaces but eventually collecting on horizontal surfaces in crevices at the lowest areas in the chamber. Generally, the vapour deposit is not tenacious and can be easily brushed off after welding, but precautions must be taken to prevent the droplets and spatter falling onto moving parts in fixtures and manipulators, particularly trackways. Simple shielding is the practical answer. Bearings can also be made self-clearing and, whenever possible, trackways should not present a horizontal level surface on which spatter may collect.

Wire feed

The importance of fit-up between parts to be welded has been emphasized repeatedly. The degree of tolerance on fit-up between the two parts is essentially a function of two factors: the part thickness and the physical properties of the molten material. The surface tension of the liquid metal determines the stability of the weld pool between the abutting faces: if the surface tension is low, there is a tendency for the pool to run out of the gap, and this is aggravated if the gap is wide, i.e., if the fit-up is poor.

The recent advent of wire feed in conjunction with EBW could do much to reduce the dimensional accuracy normally required in machining and fixturing, since the surface undercut can be compensated for by the addition of filler wire.

The benefits of wire feed will be particularly appreciated in deep-penetration welding where the attainment of good fit-ups is both technically difficult and expensive. For welding really thick plate by electron beam, a positive gap of wide tolerance can be used, though there are practical difficulties in applying this technique. The electron beam is focused into the gap and wire is fed into it; the wire melts and brazes or welds to the sides. Insufficient data are available on this new method, but it is considered to have most promising possibilities. Beams of low intensity can be used since the beam is no longer required to cut its way through the material. The heat from the beam is used more efficiently, i.e., the energy absorbed per unit volume of weld is reduced, lower temperature gradients are experienced, and there is less tendency for voids to develop, as there is more opportunity for the metal to release any of its dissolved gases. Feed material of different quality or alloy composition may also be used.

It may be readily imagined that wire feed has also proved to be useful in building up worn parts or reclaiming parts which would otherwise be scrapped because of machining errors, porosity, or cracks. Conventional wire-feed equipment is unsuitable for direct application to EBW, since high positional accuracy is required between the wire, the beam, and the joint interface. Since manual control inside the vacuum chambers used in EBW can be difficult, some form of remote control is usually necessary to maintain the present parameters of the system and also to ensure correct wire feed rate. Wire-feed equipment of a type suitable for the more stringent requirements of EBW is commercially available and is fitted with a variable-speed drive and tachometer indicator (see chapter 4).

Pumping equipment

The selection of pumping equipment for EBW machines is, in the main, for the machine designer to decide; the user is rarely involved. Occasionally, however, existing pumping arrangements must be modified to decrease the evacuation time, or simply to increase pumping capacity to cope with extensions made to chamber volume. In these cases, a basic knowledge of typical pump systems is essential.

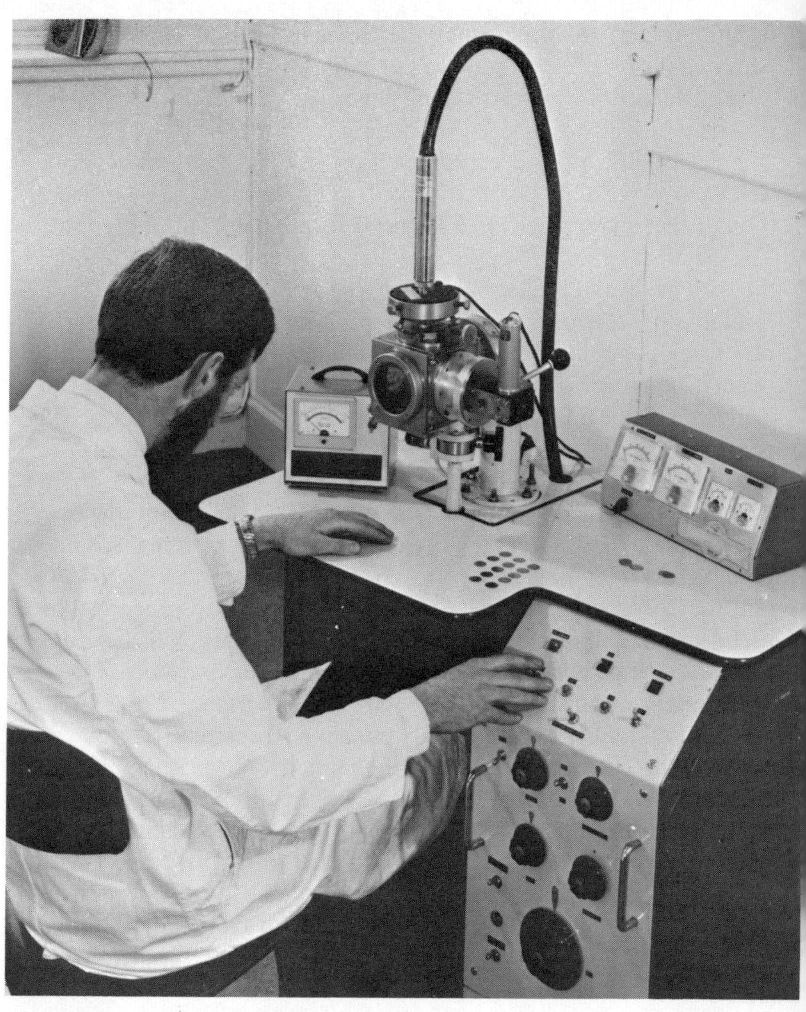

Fig. 8.3 'Mini' beam welding machine suitable for capsule welding and similar applications (*Courtesy Sciaky Electric Welding Machines Ltd*)

The operating pressure of vacuum EBW machines is approximately 10^{-4} torr. For chambers up to 35ft^3 (1000 1) in volume, evacuation times of less than five minutes can be achieved comfortably with a rotary pump backing a single oil-diffusion pump, the rotary pump taking the system from atmospheric pressure to less than 1 torr in about half the time, the diffusion pump the rest of the way in the remaining time. Pump-down times of a few seconds are possible, but this is normally necessary only in certain repetitive operations. Figure 8.3 shows a very small system for individual components which has a pump-down time of 30s without a prohibitive initial cost. As the chamber size increases, the size of the rotary and diffusion pump required to evacuate the chamber in such a short time escalates to large dimensions; so does the cost. The problem must then be approached in a different way.

An examination of the characteristics of conventional mechanical rotary pumps and diffusion pumps will reveal that each has its optimum performance in a clearly defined range (see Fig. 8.4). The

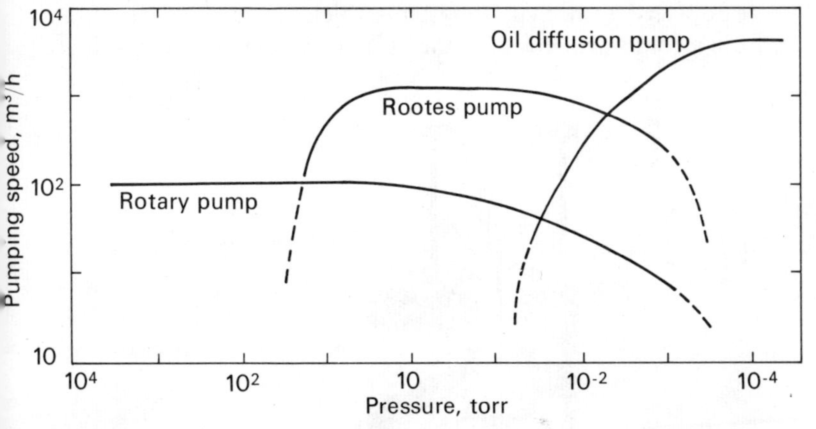

Fig. 8.4 Regime of operation of vacuum pumps (*Courtesy GEC Power Engineering Ltd*)

two regimes do not adequately cover the range between about 1 torr and 0·1 torr. By adding a third type of pump to the system, a booster of either mechanical (Rootes) or vapour type, the gap can be closed. In fact the Rootes-type pump has its optimum performance between 10^{-2} and 10 torr, with useful capacity to boost the rotary pump at 100 torr, or even at atmospheric pressure, while the vapour-type has a useful range extending from about 10^{-3} to 1 torr. Either of these pumps adds substantially to the cost of the pumping system, but may be justified on economic grounds.

Evidence is now available to show that, provided the electron beam is generated and formed at low pressure, i.e., 10^{-4} to 10^{-5} torr, it can be brought into a region operating at a substantially poorer vacuum, i.e., at a pressure of up to about 10^{-1} torr, and welds made which are very similar to those obtained at lower pressures. This affords an opportunity to retain several advantages such as the inherent cleanness associated with a vacuum, deep penetration, and minimum heat input, while at the same time offering a shorter pump-down time, a simpler pumping system, or both. The ability to work at pressures around 10^{-1} torr may, therefore, see radical changes in pumping systems currently used. The oil-diffusion pump with its inherent disadvantages and attendant complications, e.g., long heating-up time, low efficiency, large bore connections, valving, and cumbersome construction, may in some installations be displaced by mechanical systems which can quickly reach 0·1 torr from a cold start. However, the beam-forming region will still need

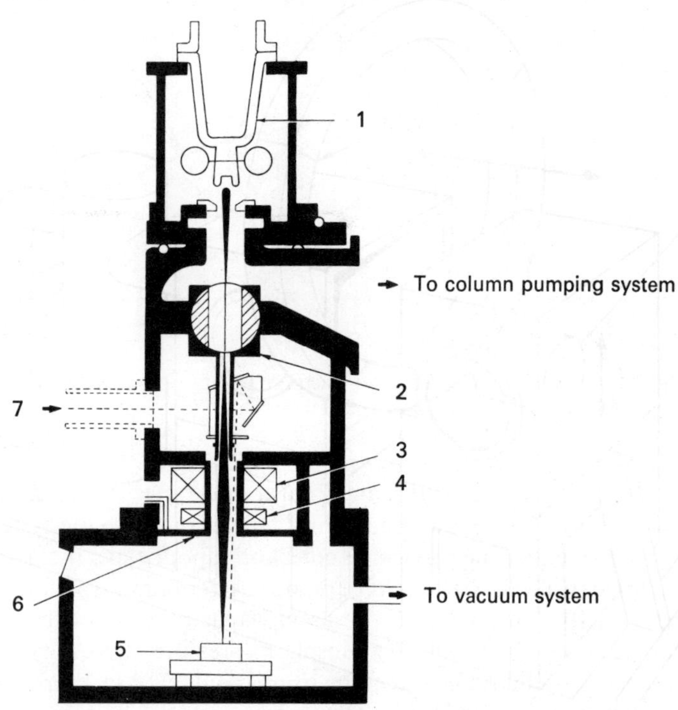

Fig. 8.5 Soft-vacuum plant: 1—electron gun; 2—column valve; 3—magnetic lens; 4—deflection coil; 5—workpiece; 6—water-cooled heat shield; 7—optical viewing system (*Courtesy Hawker Siddeley Dynamics, Electron Beam Division*)

a hard vacuum, and hence the use of a small oil-diffusion pump. An example of this system is illustrated in Fig. 8.5.

Special considerations may be necessary with very large assemblies. In such circumstances, the costs of massive vacuum chambers to encompass the entire job may be prohibitive, and recourse must be made to alternative techniques. Many ingenious and more or less practical solutions have been proposed to provide adequate vacuum locally in the immediate vicinity of the joint. This principle can be seen in Fig. 8.6, which illustrates a method for the assembly of a large ring. More complex sections could be sealed by using inflatable seals. A further example is the welding of a large diameter tube, shown schematically in Fig. 8.7. Inner sealing is accomplished

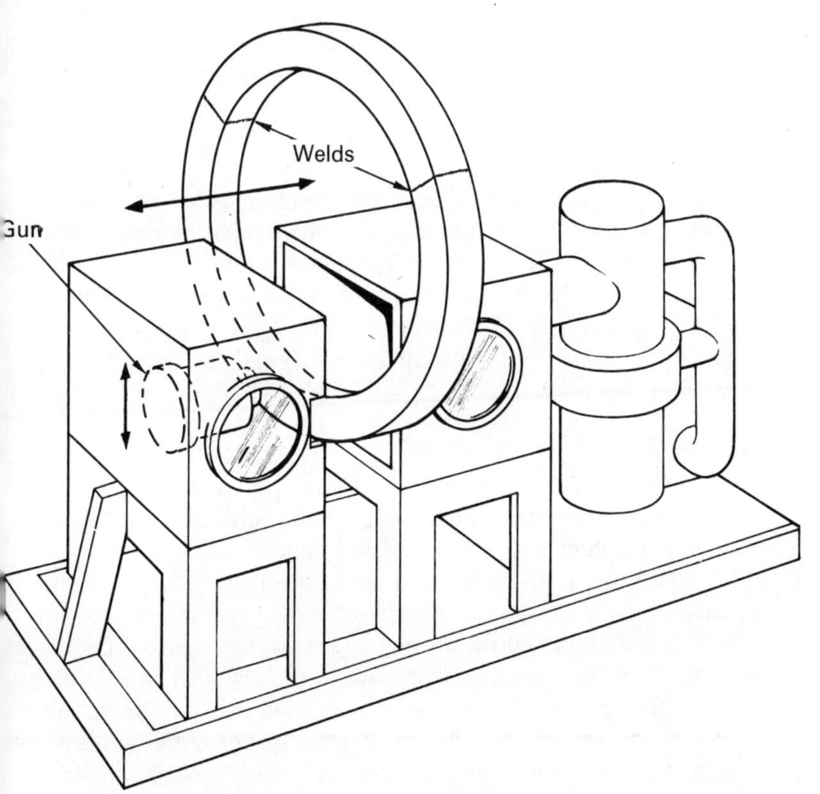

Fig. 8.6 Welding large rings without vacuum chambers (*Courtesy Sciaky Electric Welding Machines Ltd*)

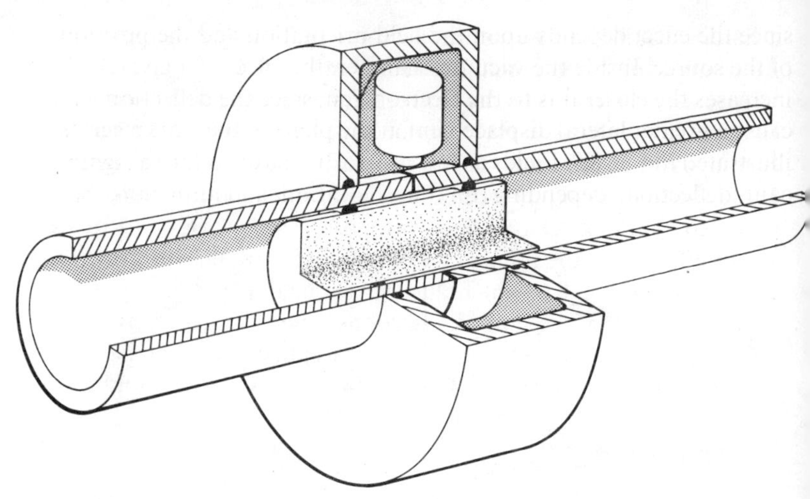

Fig. 8.7 Welding large-diameter tube without vacuum chamber

by an inflatable ring mounted on a shaft which is slid inside the large tube. The outer seal supports a chamber and electron gun. The motion required for welding can be produced either by rotating the gun inside the chamber and around the tube or by rotating the chamber and fixed gun on the seal around the tube.

Walking seals have been proposed, but, though ingenious in principle, become extremely difficult to engineer in practice. They should not be dismissed for all applications, but it is doubtful whether they will find general use.

Magnetic fields

The very low inertia of an electron beam causes it to be readily deflected by a magnetic field. Stray magnetic fields can have a most undesirable influence on beam stability and precautions must be taken to eliminate them. Special attention must be given to the location of EBW equipment in order to avoid siting it near to other plant which might produce strong magnetic fields. Jigs and fixtures and other ancillary equipment placed in the chamber must either be demagnetized or manufactured from non-magnetic materials. Electric motors inside the vacuum chamber can produce significant fields and should be magnetically shielded from the beam. Parts for welding should be demagnetized where relevant. It is not possible to give typical figures for magnetic fields which may be significant, since the effect depends upon the field orientation and the position of the source. Inside the vacuum chamber, the effect of a given field increases the closer it is to the electron gun, since the deflection will cause a greater lateral displacement at the plane of the workpiece as illustrated in Fig. 8.8. Fields as low as 2 to 3G may produce a significant deflection, depending on their location and direction, but

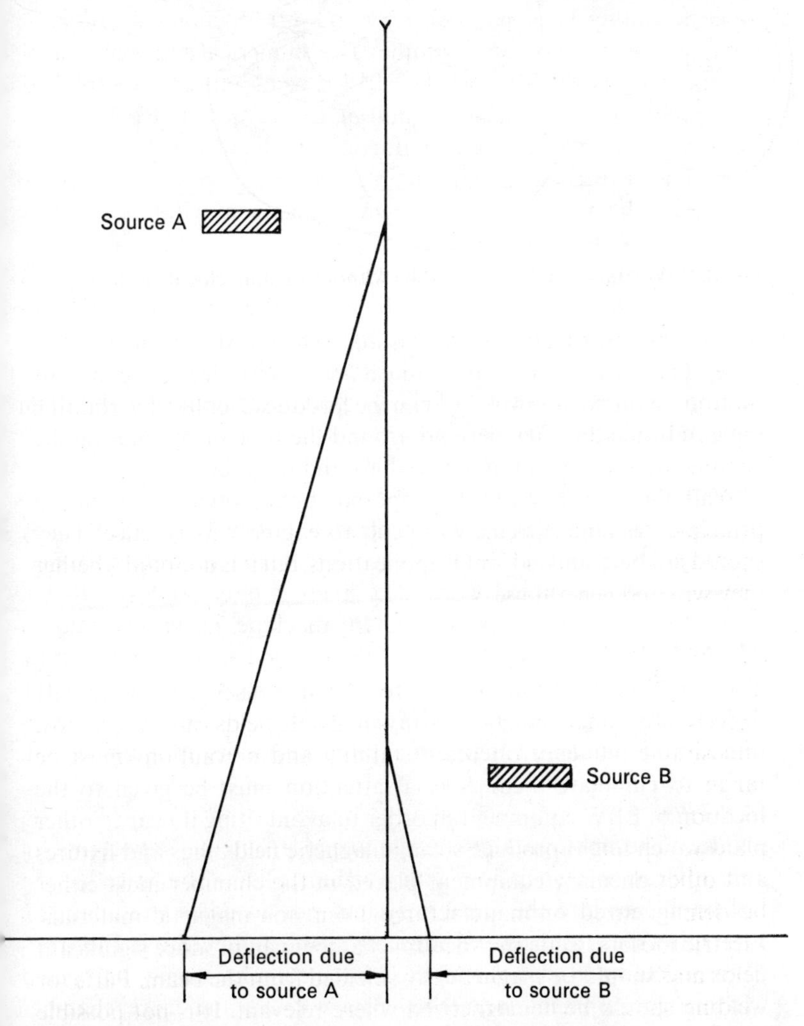

Fig. 8.8 Effect of position of magnetic source on beam deflection

considerable work has yet to be done before a truly quantitative assessment of the potential hazard of magnetic fields can be reached. Where demagnetization is not possible, recourse can be made to shielding by mu-metal.

Automatic control

The application of EBW to the rapid production of identical parts has inevitably been paralleled by the introduction of automatic manipulating devices and, in some cases, numerical and tape control, particularly in the USA where EBW is being more fully exploited. The potential advantages of automatic control in EBW are an increase in throughput and a reduction in reliance on the skill of the operator. Precise and reproducible action is a prime requisite to maintain the narrow electron beam in alignment with the joint line throughout the welding sequence.

In view of the high capital cost of EBW equipment, the opportunity of increased throughput offered by automation becomes extremely attractive. To justify the expense inevitably involved in the use of very accurate drives in conjunction with high-speed motors, automation is, however, likely to prove economical only when applied to large-scale mass production.

Maintenance and equipment layout

Equipment failure is always expensive, particularly when costly plant such as that for EBW is involved. There are also the repercussions of interrupted work. The plant is often so costly that it precludes the availability of a standby machine. Ensuring a smooth rapid flow of work into and out of the welding area is a prerequisite of economical operation of plant and it is thus necessary for the production engineer to have an understanding of the problems involved in the maintenance of equipment and its optimum layout.

Cleanliness

Vacuum technologists are quick to point out the importance of cleanliness in association with vacuum plant. The introduction of impurities into the vacuum system can do much to impair the pumping speed and general efficiency of the pumping unit. Electron-beam welding machines with their often complex manipulating and fixturing equipment, numerous rotary seals, and large chamber surface areas present special problems to the vacuum technologist.

Intricate handling equipment and associated gearing can offer innumerable dust traps, and regular cleaning is vital to ensure efficiency of pumping. When the chamber is open to atmosphere, the large, cold interior walls become effective condensation surfaces for water vapour and the problem can become acute if the environmental humidity is high. Not all chamber walls are identical; some manufacturers use stainless steel and others mild steel; epoxy resin coatings have also been used. Different surfaces act differently as condensation faces, but if the condensation of water presents a real problem, one simple solution is to introduce a tray of phosphorus pentoxide or other rapid moisture-absorber into the chamber. On the premise that prevention is better than cure, however, some precautions taken during the installation stages can pay dividends later. The ideal position would be an operating theatre environment, but this is clearly not a practical proposition. About the worst is a location in an open general welding shop. A compromise arrangement is a section in a clean, dry part of the factory such as a toolroom, final assembly shop, or inspection bay.

Cleaning parts and preliminary setting-up operations should be carried out away from the immediate welding area and the chamber left open to atmosphere for minimal periods.

Accidental contamination of the electron-optical system can cause embarrassing delays since often it is then necessary to dismantle the whole electron column, clean the component parts, reassemble and re-align. The whole sequence can easily take an hour. Partly to reduce this risk and partly to reduce the pumping time, shut-off valves are now introduced between the electron gun and the vacuum chamber so that, except during replacement of filaments and general maintenance of the column, the unit is kept under vacuum and atmospheric contamination is thus minimized.

Maintenance

In general, technical personnel with electronics experience appear to be more suited to the maintenance requirements of an EBW plant. Ideally, a combination of experience in pumping systems, instrumentation, high-voltage circuitry, and precision engineering is required and, although the services of staff trained in each particular field can possibly be drawn upon from outside the EBW section, it is of obvious advantage to train staff specially so that on-the-spot attention can be given to occasional breakdowns of which there is always a high incidence in the early days of installation. It is well outside the scope of this book to present maintenance programmes for all the types of machine now available, but many of the problems are common to all. Preventive maintenance becomes increasingly important as the complexity of the plant increases. There is little doubt that downtime is minimized if a regular programme of inspection and regular maintenance is conscientiously pursued. Maintenance of electron-beam equipment can conveniently be classified into its basic component sections:

electron-optical column and accessories;
high-voltage power-supply circuit;
ancillary supplies and controls;
vacuum system and vacuum pumps;
manipulators.

Simple but regular inspection of the rotary and diffusion pumps will usually be sufficient to ensure satisfactory operation throughout their working life. Oils should be carefully maintained at the specified levels and the diffusion pump should be cleaned internally once every 2000 hours' running and new oil added. Routine conventional maintenance of rotary pump motor and drive belts should be planned. Vibration transmitted between the rotary pump and the vacuum chamber is minimized by mounting the pump on anti-vibration feet and by inserting a convoluted vibration-damping pipe at the rotary pump inlet port. Although no trouble should be experienced here, occasional inspection is recommended. Pressure-measuring instruments should be periodically checked against a standard and the gauge heads cleaned as recommended by the manufacturers at, say, six-monthly intervals. A wide variety of vacuum valves is now used in connection with different EBW machines, but generally they will not require any special attention. The plate valve, sealing the diffusion pump from the chamber, seats on an 'O' ring and the latter should be kept clean by occasional careful wiping; it is possible that excessive spattering during welding can lead to metal droplets being deposited on this ring. A poor seal at this point can result in the hot diffusion-pump oil being exposed to atmospheric pressure. This will produce 'cracking' of the oil and render it useless as a pumping medium.

A scrupulously clean electron-optical system is essential to ensure satisfactory performance of the gun. Component parts of the cathode system should be cleaned in Analar reagents and subsequently handled by personnel wearing clean gloves. Any scratches on polished parts of the cathode unit can seriously impair performance and should be removed. Optical alignment systems have a glass shield at the chamber end to protect the lenses from vapour deposits. Metal vapour condenses on this glass shield and will eventually reduce the amount of light passing through it, thus darkening the field of vision. At this stage, the glass must be removed and cleaned; aqua regia is a common cleaning fluid. The operation of removing and cleaning the cover glass is a simple one; leaving it in too long will impair operator observation and may lead to faulty alignment.

Electron guns are often water-cooled through capillary tubing. Blockage of these tubes can result in serious damage. A clean supply of cooling water is essential.

The power-supply system is obviously an important part of an EBW machine. Bends in the supply cable of high-voltage machines should be of as large a radius as possible and movement limited to essential operations such as gun overhaul. Transformer oil level should be inspected regularly and topped up when necessary.

Maintenance of component parts of the vacuum chamber and manipulating equipment is fairly straightforward. Motor brushes should be inspected and electrical contacts, particularly relay mechanisms, should be examined for wear at intervals. Interior lamps and observation windows will have to be removed when vapour deposition affects viewing, and cleaned in the same way as glass shields in the optical alignment system. Lubrication at low pressures can present problems, particularly when loads are high. It will then be necessary to employ special lubricants.

Electron-beam welding equipment produces secondary emission in the form of X-radiation and some chambers need to be lead-covered to minimize the potential hazard. If lead covers have to be removed for any reason, it is obviously essential to replace them accurately. A radiation test should be carried out after reassembly in the interest of safety. Relevant radiation regulations should be applied, such as the wearing of radiation badges and the occasional blood count.

All vacuum seals must be kept clean and those regularly opened, such as the main door seal, should be carefully examined at frequent intervals. The refusal of a pumping unit to evacuate a chamber can be attributed to one or both of two causes: a leak or high outgassing from inside the chamber.

Outgassing is caused by the presence of materials with high vapour pressure in the vacuum chamber. Care must obviously be taken to ensure that such materials, e.g., wood, oil, grease, and water, are not introduced accidentally. Some outgassing is inevitable with the type of equipment used in EBW machines, but there is a limit to what can be accepted. A relatively easy check can be made occasionally as follows. Evacuate the chamber to a pressure of 10^{-4} torr, and isolate it by closing the high-vacuum valve. By a knowledge of the chamber volume in litres, calculate the 'leak rate' or 'outgassing rate' in torr-1/s. As an example, a chamber of volume 1000 1 shows a pressure rise from 1×10^{-4} to 1×10^{-3} torr in 60s. The 'leak rate' is then $(9 \times 10^{-4} \times 1000)/60 = 1{\cdot}5 \times 10^{-2}$ torr-l/s. A figure of less than 3×10^{-2} torr-l/s is usually acceptable. If the figure rises much above this level, an examination for leaking seals and possible materials with low vapour pressure should be undertaken. Leak-detection equipment is available, but specialist knowledge is essential for its satisfactory operation and an approach to the manufacturers or other vacuum experts is recommended.

Spares for EBW machines can present a problem for the production engineer. In many cases, spares can be obtained overnight from the manufacturers, but obviously efficiency of this kind differs from one manufacturer to another. It is probably not economical to stock an entire selection of replacement parts even when taking into consideration the downtime involved in awaiting delivery. Several items should be regarded as essential, others depend upon estimated delivery times. Components such as filaments, 'O' rings and other seals, viewing lamps, electrical contacts and motor brushes, and fuses, should be regarded as essential. Most other parts are either very expensive or capable of temporary repair pending delivery from manufacturers. It is gratifying to note that wear and tear in electron-beam equipment is quite low and the plant should continue to give good service for ten or even twenty years.

Plant layout

A full study of plant layout principles is outside the scope of this book and the reader is referred to the many detailed texts available on the subject. Electron-beam welding is in some instances a special case and reference will be made to some of the major considerations.

Previous considerations of cleanliness and maintenance have shown that a smooth flow of prepared work in a reasonably clean dry environment is desirable for optimum running of an EBW facility. Layout of plant should be such that this work flow is not impeded. Two separate working surfaces should be provided, one for general fitting work, the other for final setting-up and cleaning operations. Storage of work before welding can be an important aspect. Often it is desirable to clean large numbers of components at a time; clean storage space must be provided so that the parts can be extracted from stock and inserted directly into the fixturing device. Heavy parts may require the services of a crane, and allowance should be made to install such lifting equipment if there is a possibility of having to handle large items. The layout of the machine itself should allow for future requirements; for example, it may be necessary to increase the chamber size by adding extension sections to the existing machine. The popularity of EBW is undoubtedly on the increase; it would be sound policy to bear in mind the possibility of acquiring more machines in the future by selecting a site with potential for expansion.

Quality control and inspection

Quality control encompasses all techniques used to reduce or to eliminate the causes of defective production. Inspection serves as an important reference for quality control, but by itself it can never be a remedy for substandard work.

The technology of quality control, namely the techniques used in assessing how a product should be studied during its passage along a production line, has been discussed elsewhere in detail by many authorities; the reader will also be referred to the many different techniques of sampling for inspection. The present discussion will be confined to an outline of the specific factors which can influence the efficiency of EBW.

Techniques of quality control can be divided conveniently into three groups:

control of components before reaching the processing stage;
control during processing;
control after the operation.

We shall consider preweld quality control first. Compositional variation in the material being welded can lead to variations in weld integrity and strength; weld porosity can be introduced by the inclusion of gases or elements with high vapour pressure into the material. The problems of alignment and dimensional accuracy of parts for EBW have been discussed earlier. In general, abutting faces must be in close contact and often this calls for high dimensional accuracy if weld defects are to be avoided. The small assembly shown in Fig. 8.9 has two circumferential welds, one at the top the other at the bottom. It was discovered that, unless the fit-up between the bobbin lands and the internal wall of the spool was better than 0·001in. (0·025mm), considerable barrelling took place at the welds because of shrinkage, and that severe undercutting at the surface occurred if the fit-up was allowed to exceed 0·002in. (0·05mm). Such defects are aggravated by poor surface finish on the abutting faces.

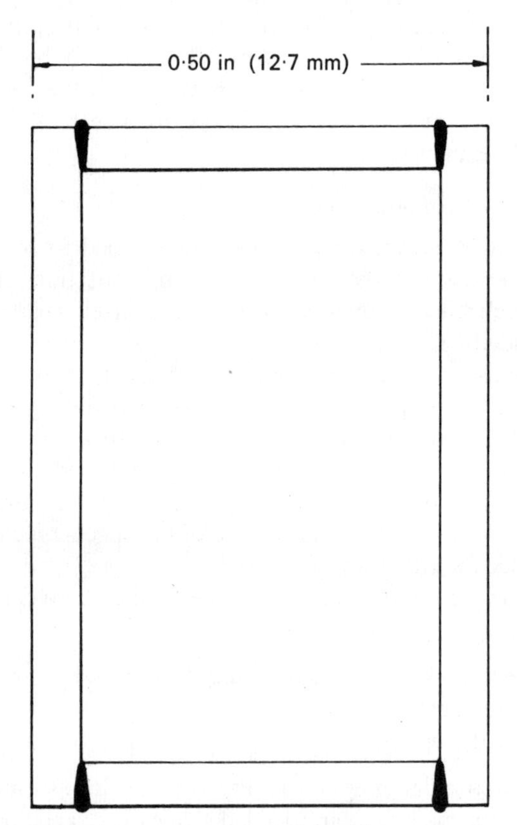

Fig. 8.9 Assembly with two circumferential welds

During the assembly and welding stages, quality control is principally concerned with cleanness, both of parts and of equipment, and with optimum performance of machinery, including jigs and fixtures. The significance of all these has been discussed previously and the importance of cleanness is particularly emphasized.

After the welding operation, the parts will be inspected. A decision must be taken at this stage as to which inspection technique to use and which sampling procedure to follow. In welding production batches of components, an assessment of the suitability of any inspection technique must be carried out. Generally, the assessment consists of correlating data collected from metallographic examination of sections with information obtained by various methods of non-destructive testing, and then setting an acceptance standard which will allow a variation about an optimum. The acceptance standard is, therefore, dependent on metallographic examination and its correlation with the mechanical strength. It will also be closely connected with repeatability of equipment settings. Metallographic examination of joints will reveal any tendency of the material to crack or to produce porous welds, but it is important to investigate the joints typical of a final product and not of an oversimplified simulation, since joint design can greatly influence weld soundness. Mechanical testing of welds will indicate the strength of the joint, but, again, care must be taken to ensure that typical joints are examined. For example, if the weld is not to be machined after welding, testing should be carried out on unmachined welds; this is particularly significant in fatigue testing. Mechanical testing can also reveal whether or not postweld heat treatment will be necessary to improve the mechanical properties of the joint.

In many cases, a comprehensive investigation of weld properties is not necessary. The EBW process may be required only for very lightly loaded parts or for small sealing welds. In these circumstances, quality control of EBW becomes relatively easy.

Many inspection techniques are available of course, but the following have been successfully applied to EBW:

surface defects	dye penetrants
	fluorescent detectors
internal defects	magnetic crack detection
	radiography
	colour radiography
	ultrasonics

There are five basic possible defects in electron-beam welds:

porosity;
lack of sidewall fusion;
lack of weld penetration;
internal cracking;
surface defects, e.g., undercut and surface cracks.

Techniques of surface examination are well established and are capable of resolving most types of surface defect. Dye penetrant and fluorescent techniques are both useful; fluorescent examination is generally more efficient, but cleaner surfaces must be provided than for dye penetrant examination. Detection methods based on magnetic fields can be used, but the field direction should preferably be varied during examination to cover the eventuality of cracks or other line defects running in the same direction as the magnetic field.

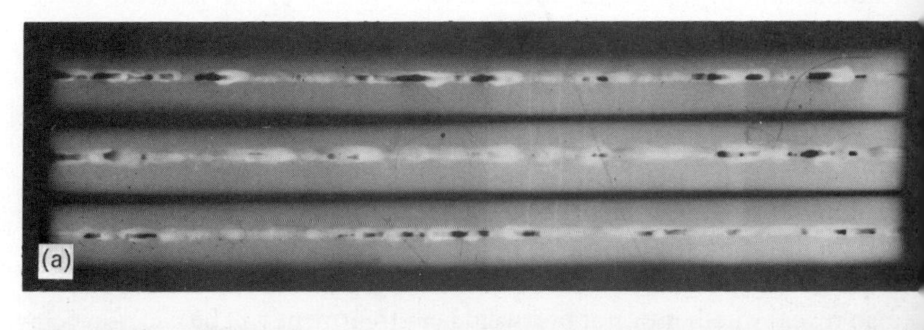

Fig. 8.10 Application of radiography to EBW inspection: (a) radiographed in plane of weld, porosity just revealed, (b) radiographed normal to plane of weld, porosity quite evident, and (c) section of weld (*Courtesy GEC Power Engineering Ltd*)

In comparison with conventional welds, electron-beam welds are in general small and alignment of the X-ray primary beam with the weld is more critical when viewing down the joint line. Radiography is particularly effective when viewing normal to the weld depth, since porosity is easily revealed, but joint design and component geometry

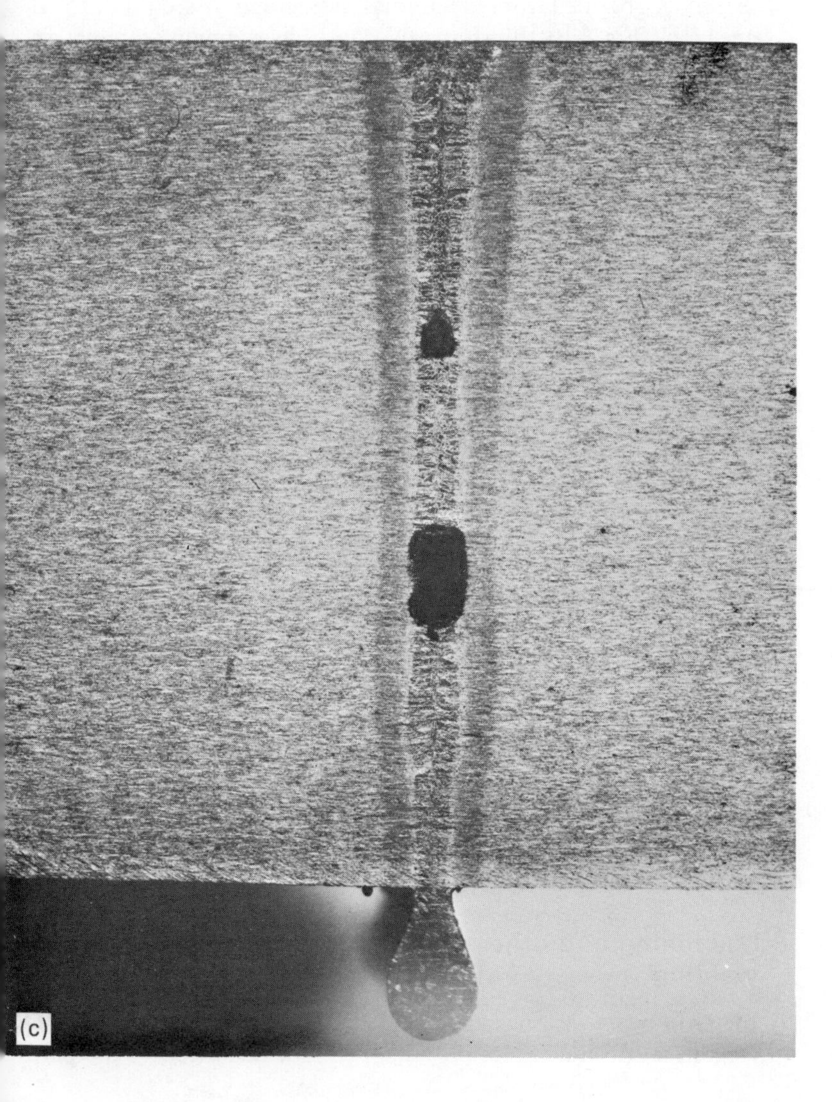

(c)

often preclude this type of examination. The effectiveness of examination normal to the weld path is revealed by examining Fig. 8.10. Three 0·50in. (13mm) square blocks, each with a longitudinal weld, were radiographed in the plane of the welds (Fig. 8.10a) and normal to the welds (Fig. 8.10b). It is clear that, although the presence of porosity is revealed by viewing down the weld, a more revealing picture is obtained from the normal radiograph. A section through one block (Fig. 8.10c) illustrates the type of defect under examination. Under optimum radiographic conditions, defects caused by porosity, lack of fusion or of weld penetration, can be detected.

The technique of colour radiography is still in an early phase of development. Currently available films have low speeds so they are suited to thicknesses of up to 0·4in. (10mm). The principal advantages of colour radiography are:

the human eye can detect changes in colour more readily than changes in grey tones;

resolution is increased since there is significantly less scattering effect.

Ultrasonic inspection can be more versatile than radiography since the angle of the incident beam can be continuously altered during examination which offers a higher probability of defect detection. Typical ultrasonic traces of two welds are shown in Fig. 8.11a and c, together with the corresponding macrosections in Fig. 8.11b and d, respectively. The component being examined here is a small piston shown schematically in Fig. 8.11e, which also illustrates the method of ultrasonic examination. Defects as small as 0·01in. (0·25mm) were readily detected and located in this application.

With both radiography and ultrasonics, material structure and composition may have a significant effect on the results. For example, alloys with large grain size, such as austenitic steels, reduce resolution with radiography. They also introduce superimposed interference on the ultrasonic image.

Machining and other preparatory operations before welding

Electron-beam welding has acquired for itself the reputation that it is capable of producing fusion welds without distortion of the part.

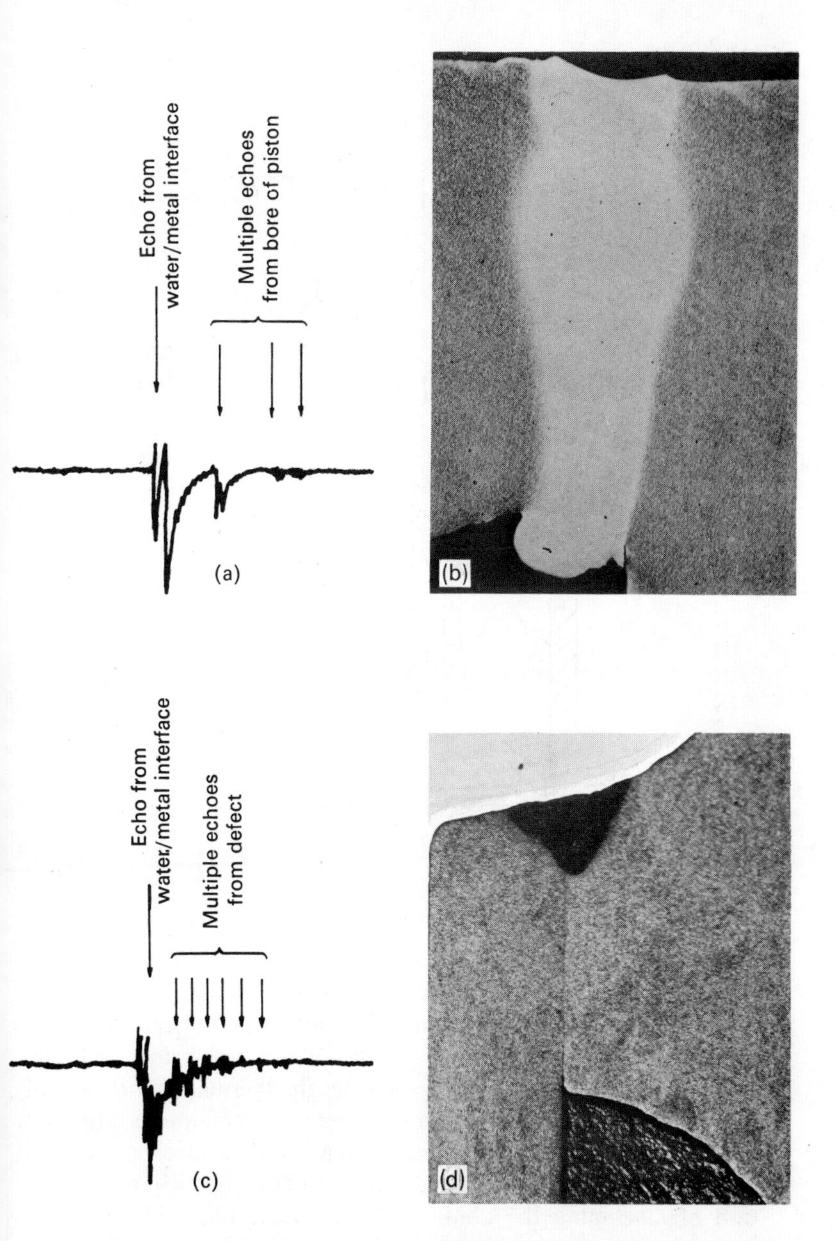

Fig. 8.11 Application of ultrasonics to EBW inspection: (a) and (c) are ultrasonic traces corresponding to macrosections (b) and (d); (e) (overleaf) shows piston under examination (*Courtesy GEC Power Engineering Ltd*)

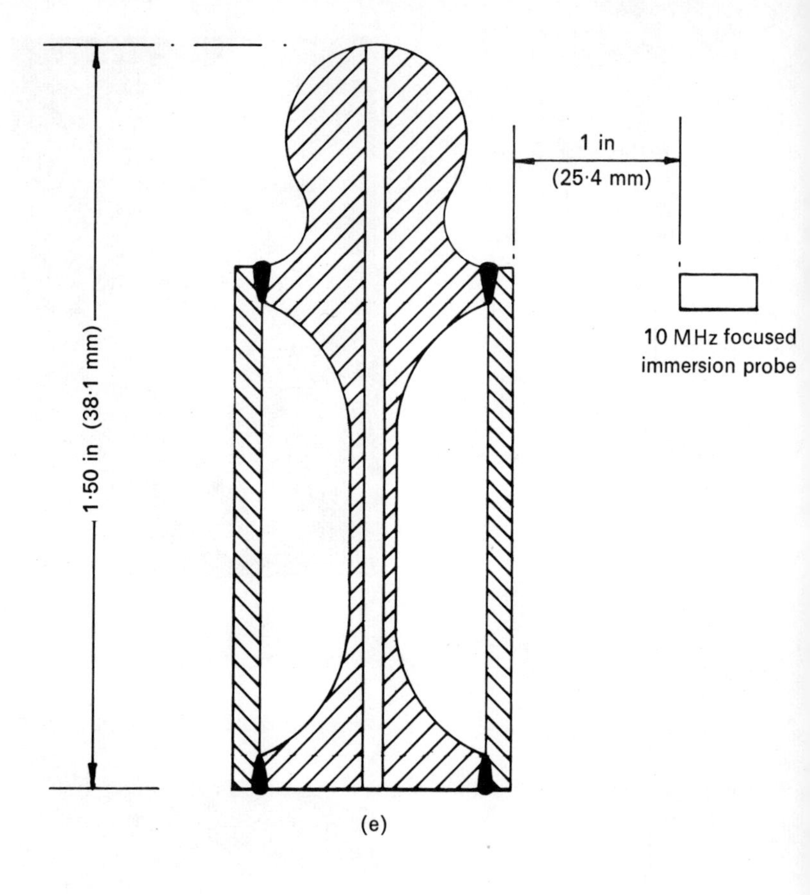

(e)

Although this may be practically true for some configurations, a measure of distortion is always present. In most applications, the distortion is so small as to be insignificant, but in others it must be taken into account in the machining of the part before welding. Naturally, the finer the weld the smaller the thermal distortion and shrinkage displacement. However, there are limitations imposed not only by the electron-optical system itself but also by the geometrical arrangement of the workpiece and the practical consideration of machining the abutting edges to the required degree of accuracy. Even so, there are problems associated with excessively narrow welds such as root porosity, irregular penetration, and the risk of missing the joint line.

The extent of the thermal distortion of the part is related to its own rigidity, the rigidity of the holding fixtures, and the configuration of the weld relative to that of the part. For example, in a planetary gear weld, located at the web, a reduction in diameter of 0·2 per cent was detected. However, concentricity, swash, dishing, and alignment of the gear teeth were held within close limits so that only the final grinding operation had to be left until after welding.

The shrinkage distortion in a circumferential weld occurs axially and can be accurately predicted, thus making it possible for allowance to be made in the machining of the component parts. In a typical example, where five rotors 12in. (305mm) dia. were welded together, a total axial shrinkage of 0·030in. (0·75mm) was detected. The thickness of each joint was 0·100in. (2·5mm). The concentricity of the rotors was held to within 0·004in. (0·1mm).

If it is established, from previous experience or from work on simulated parts, that the distortion can be tolerated or that adequate allowance can be made for it, the parts can be finish-machined before welding. It may be desirable, however, to machine either the upper or lower bead, or both. The removal of the lower bead normally results in improved fatigue properties. There is little advantage in removing an upper bead of good shape, although a second beam pass at reduced power (cosmetic pass) is often used to improve the appearance of the weld.

The surface finish of the abutting faces is a factor that does not normally present any problems. The good surfaces produced by turning, milling, or grinding are quite adequate. Typical values producing good welds fall in the range between 50 and 75 $\times$ 10^{-6}in. (1·25–2μ). However, gaps between the abutting surfaces have to be closely controlled. In practice, gaps of up to 0·004in. (0·1mm) can be tolerated over short distances along the weld line, but the finer the beam the narrower the permissible gap, and it is advisable not to exceed 10 per cent of the beam diameter. When there are wider gaps, it is feasible to produce satisfactory welds by the application of automatic wire feed, although considerable experience is required. In an extreme case, it has been found possible to fill a gap 0·028in. (0·7mm) wide when welding material 0·064in. (1·6mm) thick, by feeding in wire 0·036in. (0·9mm) dia.

Welding sheet metal parts by EBW presents certain problems in preweld preparation. Generally, blanked or guillotined edges do not produce the required accuracy. In most cases the edges must be machined or ground before welding, not only to improve the dimensional accuracy but also to eliminate burrs and torn edges.

It is not sufficient to produce edges of acceptable surface finish with good abutment; the joint line itself must be geometrically accurate. This requirement is essential in guaranteeing that the joint line falls below the beam during the welding operation. This machining accuracy has to be considered in conjunction with the accuracy of the manipulator, as the errors from both sources are cumulative.

Cleaning

Experience by many users has shown that component-cleaning requirements for EBW are more stringent than in other fusion-welding methods. Firstly, there is the general cleaning of the part. This is advisable because of the vacuum environment; foreign matter or excessive rust or scale will tend to absorb moisture or oil which will degas during chamber evacuation, leading to extended pump-down times.

Cleanness in the vicinity of the weld and of the abutment faces themselves is even more critical. Gross oxide particles, for example, may become dislodged during welding and lead to inclusions in the weld. It has also been observed that the dissolution of surface films on the abutment faces during welding results in a degradation of the metallurgical structure of the weld and a deterioration in its mechanical properties. In certain cases, such as the welding of titanium and aluminium, gross porosity may occur in the weld. The edges are often pickled, but the removal of a thin surface layer by dry mechanical means is probably better. It is advisable to degrease the part immediately before welding; also, the weld zone should be washed with a suitable solvent such as acetone. Care should be taken to avoid touching or contaminating the edges after this treatment.

Economic considerations

Methods of cost analysis are well known in the engineering industry and are applied in various degrees of sophistication. Reference can be made to numerous works on the subject and no attempt will be made here to deal with cost analysis in a wider sense; rather, those cost factors that are of more special significance to EBW will be discussed.

As indicated earlier, the cost of producing a satisfactory EB weld will include the cost of such operations as the preparation of the material before welding, the welding operation itself, heat treatment after welding, the extent and frequency of inspection to achieve the required reliability, the cost of scrap, and the cost of salvage. All ancillary operations are common to both EBW and other fusion-welding processes and the production engineer can apply his knowledge of the cost of such operations to EBW. However, the actual welding operation itself is radically different and will, therefore, be examined in some detail.

Many factors contribute to the cost of producing an EB weld, and users, depending on the nature of their work, will place different emphasis on different factors. For example, jobbing firms will need a continuity of work and will prefer not to interrupt for experimental work. They do not normally have a large technical supporting staff to their operations. By contrast, an organization developing its own product may allow a proportion of the plant's time for experimental work, thus reducing the hours available for production. They may also employ qualified staff to deal with metallurgical, testing, and technique developments, increasing the overhead costs. Nevertheless, the principles applied are the same and these will now be outlined.

The machine hour rate

Because the capital cost is a major factor in EBW, it is convenient to make use of an intermediate yardstick—the machine hour rate—which says nothing in itself about output but shows in a practical way the effect of all the cost factors, direct and indirect, and the time that the facility is available. Coupling this with the output per hour enables us to derive a cost per weld. The cost factors, the productive time available, and the output per hour, will be discussed in turn.

Cost factors

FIXED DIRECT CHARGES. The decision to acquire and install a major plant is normally a matter of policy, influenced by, for example, the need to expand facilities to meet increased demand. Once purchased, the plant gives rise to fixed charges in two ways which are worked out on an annual basis:

(a) An accumulating reserve to be set aside eventually to replace the equipment; better known as depreciation or amortization (Fig. 8.12).
(b) The interest charged on the capital outlay (or that portion of it) as yet not reserved under (a) above (Fig. 8.13).

Generally, these two charges represent the major proportion of the cost of a weld.

If the amount reserved is the same each year, the interest charges on the diminishing capital still outstanding will diminish in equal steps. Thus, on average, interest charges will amount to the full charges on half the outlay, assuming no fluctuations take place in the rate of interest.

However, the period over which the equipment is to be amortized must be determined with some care. In a rapidly developing technology such as EBW, the equipment may become out of date

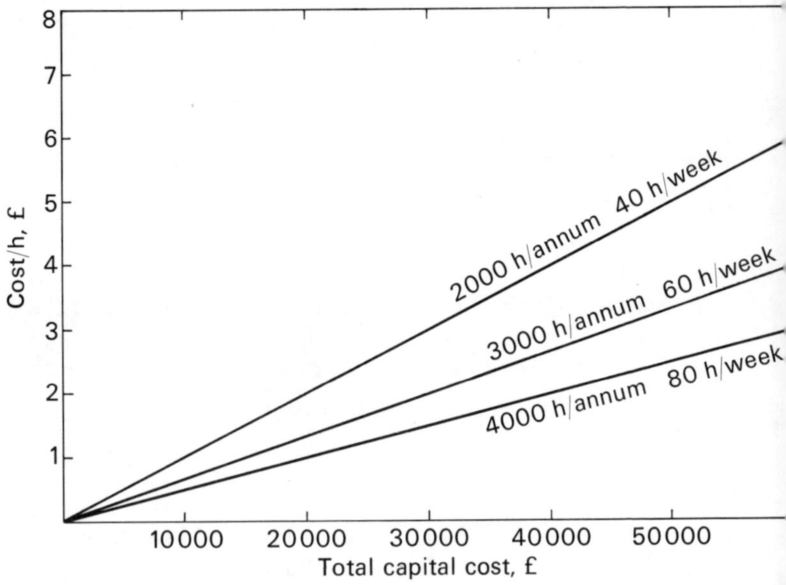

Fig. 8.12 Amortization charges

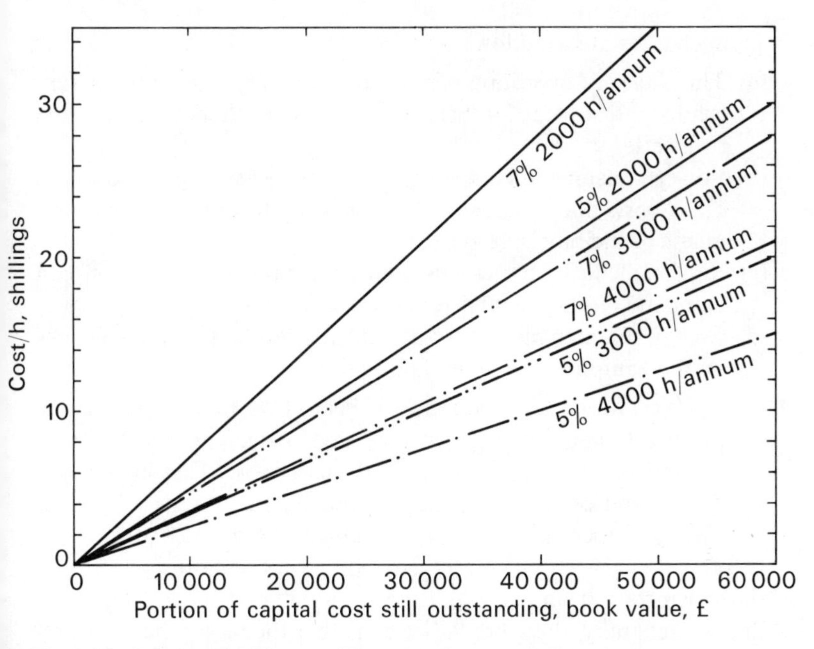

Fig. 8.13 Interest charges

because of continuing innovation. Existing equipment may become so inefficient as to require an early change. On the other hand, actual wear and tear in an EBW machine is likely to be so small that the life of the equipment may be fifteen years or more. Indeed, experience so far has shown that early equipment is still performing well even though lacking some of the innovations found in more modern equipment.

Even so, it is perhaps reasonable to amortize an EBW plant in less time than this, say in ten years. In a commercial operation, a cost accountant, in view of the risk of obsolescence, will wish to reduce this period to seven years for his peace of mind. However, he must also take care not to price himself out of the market in his natural desire to reduce the term of amortization. Figures 8.12 and 8.13 show the considerable effect that capital charges have on the total cost of the operation.

VARIABLE DIRECT CHARGES. These include charges, other than capital, that make a direct contribution to the machine hour rate. The main charges are as follows.

(a) The wages of operating personnel. Generally one man per shift is adequate, except for large machines where he may be assisted by a setter.
(b) Power consumed. This is generally small by comparison with other costs and is due, in large measure, to the power requirements of the pumping system.
(c) Replacement. Items such as oil for the pumps and electron-gun filaments require regular replacement.
(d) Cost of maintenance. It is advisable to conduct periodic checks on the equipment to avoid failure during service.

INDIRECT CHARGES. These are the overheads that are normally added to the direct charges. They represent the cost of supporting effort to maintain an efficient operation of the enterprise. They are derived from such sources as supporting labour, factory overheads, office overheads, supervision and management, and sales overheads. Overhead charges vary widely depending on the nature and size of the organization and it would be impossible to give a universal figure. Generally, there are indications that the larger the organization, the higher the overheads, although this should be reflected in a more efficient plant utilization. Overhead charges are normally applied to the machine hour rate in the same way as are the direct charges, and generally speaking will total about the same as the direct charges.

Productive time

The next step, after adding all charges, is to work out the number of hours, over a period of one year, during which the plant can be available for productive work. This is naturally dependent on:

equipment reliability;
number of shifts to be worked;
necessity for maintenance periods;
technique development;
sample work.

In an average production shop operating on a single-shift basis, maximum available time is 40h/week or 2000h per annum. This implies that the equipment must have a high degree of reliability and that all maintenance, development, and sample work is carried out outside normal production hours.

It may be possible to achieve some 70h/week on double-shift working, but probably no more than 100h on a three-shift system. The equipment then would be 'driven hard' and it is prudent to make a more generous time allowance for maintenance and downtime. The practice of some, when three-shift working is operated, is to allow the weekend for maintenance and other similar work, leaving the five working days for productive use.

The decision on whether to apply single-, double-, or three-shift system is governed by local circumstances such as the balance of workload against plant availability, and local labour activities. Again, if double- or three-shift work is employed, supporting services must be available outside the normal day shift to ensure a smooth flow of work to the EBW plant. It must be pointed out that twice as many skilled operators are required for double-shift as for single-shift working.

In a jobbing or development shop, it would be prudent to assume a shorter plant availability as there may be less certainty of a continuous flow of work throughout the life of the plant.

Having now commented on the various cost factors and the productive time available, we are in a position to derive a machine hour rate. This varies over a wide range, from less than £5 to more than £20.

Welding time

An aspect of EBW which has been emphasized in other chapters is the speed of the process as compared with other forms of fusion welding. This is certainly the case, particularly for deep-penetration welding in one EBW pass, compared with multipasses in arc welding, involving cleaning and inspection of the weld between runs. But speed of welding alone gives no true indication of floor-to-floor time, and only in exceptional circumstances is it possible even to approach the output which, at first sight, might appear possible.

There are a number of essential ancillary operations in a typical EBW cycle. Unfortunately, most of these involve the use of the EBW plant. Such operations include loading, setting, pump-down, and unloading. They tend to consume considerably more time than in conventional welding, principally because of the vacuum requirement and the precision of the welding operation itself.

Table 8.1 lists typical operations that contribute to the floor-to-floor time in general-purpose plant together with the range of

Table 8.1 Factors contributing to the total floor-to-floor time of EBW in general-purpose plant (neglecting jigging)

Contributory operations	Time required
1. Attach to manipulating device	3s to 5min
2. Load into vacuum chamber	3s to 5min
3. Position and secure	10s to 10min
4. Close door of vacuum chamber	2s to 20s
5. Initiate and complete pumping cycle	30s to 30min
6. Run up the gun, focus, and adjust position	12s to 15s
7. Initiate and carry out weld	15s to 5min
8. Ancillary electron-beam treatment, e.g., 'cosmetic pass'	5s to 3min
9. Allow to cool	0s to 2min
10. Switch off gun and close shut-off valve	2s to 10s
11. Release vacuum	2s to 30s
12. Open door	2s to 20s
13. Release jig and remove	3s to 5min

Note: for special purpose plant it is possible to reduce most of the times, some substantially.

time required for each operation. Of course, some of the ancillary operations can be carried out in parallel which will bring about some improvement in plant utilization. It would be difficult to reduce the floor-to-floor time below one minute for a single component and, of this, the pump-down time would be some 60 per cent of the total cycle time. Welding time would be no more than 10 per cent. Constant attempts are being made to reduce pump-down time by developing 'soft vacuum' techniques; there is also the obvious solution of utilizing more powerful pumps, although the time gained must be balanced against the increased capital cost of the plant.

Table 8.2 Variation in percentage of welding time and pump-down time with number of parts per load

Number of parts per load	1	2	3	4	10	100
Cycle time (s)	90	120	150	180	360	3060
Floor-to-floor time per component (s)	90	60	50	45	36	30·6
Pump-down time (per cent)	33	25	20	16·7	8·3	0·98
Welding time (per cent)	16·7	25	30	33	41·6	49
Remaining operations (per cent)	50	50	50	50	50	50

Automatic plant designed specially to produce specific components at a high rate with better plant utilization is also feasible.

The floor-to-floor time is substantially improved once multiple jigging or multiple welds are employed. Table 8.2 shows how pump-down becomes less dominant and how welding time, as a proportion of the total, rises to an acceptable level. It might, therefore, appear that it is always advisable to load the chamber with as many parts as possible, utilizing multiple fixtures. But there are other considerations to be taken into account. Multiple fixtures can be quite costly and longer times are required for their design and manufacture. The volume of production is often the deciding factor, since it may be hardly justified to use multiple fixtures for a small batch when a rapid turn-round time is required. If special fixtures have to be manufactured, their cost will be completely reflected in the cost of welding as they may not be suitable for other future work. It is therefore essential that the necessary optimization is worked out by the production engineer, who may, in any case, be limited in his choice to existing equipment, being unable, for example, to alter the dimensions of his vacuum chamber.

In a typical case, where single or multiple fixtures could be used, a circular component such as a transducer requires two radial welds some 0·3in. (8mm) apart. It is not possible to carry out the two welds consecutively by means of beam deflection. The central part is in stainless iron and the two ends are investment castings in a non-magnetic stainless iron. Alignment accuracy and register are both important. Initially, components are to be made in batches of 20, but this would rise eventually to batches of 100 about once a week. Since the manufacturer is installing new equipment for this job, the production engineer has a free choice of equipment, within the limits allowed by the economics of the operation. He is given to understand that there is likely to be more work of a similar nature in the future.

A number of alternatives are open to the production enegineer. He may choose to utilize a system capable of handling a number of components at a time, making the necessary allowance in his cost and lead-time estimates for the design and manufacture of a multiple jig which can index each transducer into position to be welded twice. Alternatively, he may examine the capabilities of a one-off jig which is used in a standard vacuum chamber or in one tailored to suit the specific requirements of the application.

Table 8.3 Floor-to-floor times and output for typical plants using alternative forms of jigging

Contributory factor	Size of chamber		
	Minimum	Medium	Large
Capital cost (£)	6450	10000	50000
Cost of jigs (£)	100	500	1000
Number per load	1	50	100
Pump-down time (s)	30	210	240
Floor-to-floor time (min)	2	63·5	104
Time taken to start production (weeks)	2	8	8
Machine hour (direct cost only) rate on a three-shift basis (£)	1·09	1·25	3·05
Time taken to complete production batches (min)			
Initial 20	40	23·5	24
Succeeding 100	200	127	104
Direct cost per component (pence)	8·75	6·35	12·7
Number per 20h day	620	1060	1150
Number per 5-day week	3100	5300	5750
Excess per week	—	2200	2650
Number after 6 weeks	18600	—	—
Number after 13 weeks	40400	37300	40400
Number after 14·5 weeks	45000	45000	48600

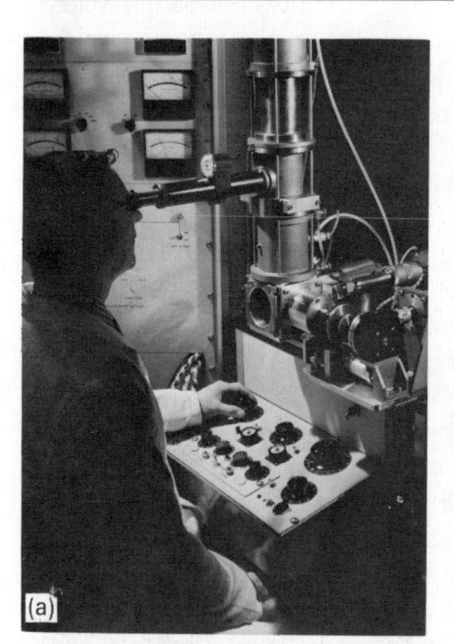

Fig. 8.14 Electron-beam welding **machines**: (a) small, for one-shot welding of circular components, (b) medium, containing multiple jig, and (c) large, with multiple jig able to take up to 100 components simultaneously (*Courtesy Cambridge Vacuum Engineering Ltd*)

An attempt has been made in Table 8.3 to compare the following three possibilities, each producing two welds per component in one minute.

(a) A small machine with two one-off jigs and a vacuum chamber tailored to suit this type of application (Fig. 8.14a), and with a floor-to-floor time of 120s, capable of 1000 components per week on a one-shift basis.

(b) A medium-size machine taking a multiple jig with two heads, capable of accepting fifty components at a time (Fig. 8.14b), with a somewhat longer pump-down time.

(c) A large machine able to take over 100 components at a time (Fig. 8.14c).

(c)

An examination of the figures shown in Table 8.3 reveals that, although the machine hour rate increases with the size of the plant, the small machine accepting only one component at a time will show an output not strikingly lower than the larger machines. It will almost certainly give a faster turn-round on the early batches and the cost per weld will also be less. It is interesting to note that the small plant will gain a head start of 18600 parts in the six weeks needed to produce the complicated jigs of the larger plant. It is only after thirteen weeks and fourteen and a half weeks for the medium and large machines, respectively, and totals of 40000 and 45000 components, that the small plant will be overtaken. As to cost per weld, the smallest plant in this case also shows up best. The major contribution to cost, as to be expected, arises from the capital and interest charges, and in the three cases examined amount to 26·6, 36, and 73·8 per cent respectively.

It must be appreciated that the above conclusions do not apply universally. It is nevertheless clear that large plant and complex

Table 8.4 Variation of machine hour rate

Cost of installation (£)		10000			50000	
Period of amortization (years)		5			5	
Interest rate (per cent)		5			5	
Shifts	1	2	3	1	2	3
Machine hours per week	40	70	100	40	70	100
Annual interest charge (£)		250				
Amount amortized annually (£)		2000				
Hourly interest rate (£)	0·125	0·073	0·05	0·625	0·365	0·25
Depreciation charge (£)	1·00	0·57	0·40	5·00	2·85	2·00
Skilled operator (£)	0·75	0·75	0·75	0·75	0·75	0·75
Water and power (£)	0·05	0·05	0·05	0·05	0·05	0·05
Total direct cost only (£)	1·925	1·443	1·25	6·425	4·015	3·05
Proportion of interest and depreciation charges (per cent)	58·5	44·7	36	87·5	80·5	73·5

Note: these figures do not include overheads.

multi-jigs, which may have to be duplicated to avoid loading delays, may not yield the best economic results, even when components are handled in large batches.

Cost of an electron-beam weld

Once the total charges, the productive hours available, and the output have been worked out, it will then be possible to arrive at the cost of each weld. Figure 8.12 shows the cost of depreciation of the

Table 8.5 Factors contributing to total cost of a welded component (electron-beam welded and others)

	Type of weld	
	Electron beam	Arc
Preparation		
Material specification (composition and gas content)	1	2
Type of material (forging—extrusion—casting)	—	—
Machining allowance	4	3
Premachining	3	3
Fitting tolerances	1	3
Cleaning	2	4
Storing	3	4
Inspection and quality control	2	3
Welding		
Handling	3	3
Jigging	2	3
Preheating	4	3
Welding	1	2
Post-heating	4	3
Finishing		
Cleaning slag and scale	4	2
Cleaning weld surface	4	3
Finish-machining	4	2
Inspection	1	1
Testing	2	2

Key: Cost contribution
1 Major 2 Normal 3 Minor 4 Negligible

plant per hour for a utilization of 2000, 3000, and 4000 hours per annum, based on a five-year amortization. Figure 8.13 shows the interest charges, as related to the portion of the capital cost still outstanding, i.e., the book value. The graphs are shown for an interest rate of 5 per cent and 7 per cent and for a utilization of 2000, 3000, and 4000 hours per annum.

In Table 8.4, a comparison is made between the direct cost per hour, excluding overheads, of two machines whose cost of installation are £10000 and £50000 respectively. The figures are based on a utilization of 40, 70, and 100 hours per week which correspond to single-, double-, and three-shift working.

Once the above calculations have been worked out, a comparison can then be made between the cost of an EB weld and a conventional fusion weld, but this must take into account the cost of preparatory and finishing operations. An attempt is made in Table 8.5 to show the relative cost contribution of each operation for both techniques. Preparatory operations for EBW are likely to cost more because of the more stringent requirements of material specifications, machining tolerances, and finishes. On the other hand, finishing operations are likely to cost considerably less; indeed, they may be non-existent in many cases.

Selected reading list

BAKISH, R. *Introduction to Electron-beam Technology,* John Wiley & Sons, NY, 1962, 452 pp.

BAKISH, R. and WHITE, S. S. *Handbook of Electron-beam Welding*, John Wiley & Sons, NY, 1964, 269pp.

FUNK, E. R. *Electron-beam Welding Symposium*, Ohio State Univ., 1967, 269 pp.

MCHENRY, H. I., COLLINS, J. C., and KEY, R. E. 'Electron-beam Welding of D6AC Steel.' *Weld. J. Res. Suppl.,* **46** (8), 1967, 337s–42s.

MELEKA, A. H. 'Electron-beam Welding as a Production Process.' *Machinery & Prod. Eng'g (London)*, **107** (2761), 1965, 796 804; **107** (2762), 1965, 852–61.

MELEKA, A. H. and ROBERTS, J. K. 'Electron-beam Welding—the Need for Further development as Revealed from Production Experience.' *Brit. Weld. J.,* **15** (1), 1968, 16–20.

SAYER, L. N. 'Quality in Electron-beam Welding.' *Brit. Weld. J.,* **14** (4), 1967, 163–9.

Procs of the 1st to 6th Electron-beam Symposia, 1959–64.

Procs of the 1st and 2nd Int'l. Confs. on Electron and Ion Beam Science and Technology, 1965–66.

Procs of the Electron and Laser Beam Symposium, 1965.

STEPHENS, T. D. *References to Electron-beam Welding,* 2nd Ed. in preparation. The Weld. Inst., 1970.

Index